Science and Philosophy - A Fresh Perspective

The Lectures

ISBN 978-1-9999664-0-9

A catalogue record for this book is available from the British Library.

Published by Merops Press
Website: www.cosmicconnections.co.uk
www.scienceandthesoul.co.uk
www.michaelpitman.co.uk

Acknowledgements: Suzanne, Marianne, Emmanuel and Françoise.

Contents

Preface

Introducing a simple structure within which to generate an internally self-consistent philosophy regarding information, physics, psychology, biology and community.

Could our scientific outlook be lop-sided? Is balance needed? Is a fresh perspective possible?

This set of lectures is intended to abbreviate an explanation of what is, basically, a fresh and simple structure - **Natural Dialectic**.

Dialectic (or dialectical method) is to-fro discourse between opposing views in order to establish reasonable truth. Of course, there may exist several antagonists in a debate but, in the case of nature, it may be shown that only polarity or, at most, trinity, holds sway.

Natural, for its part, is often construed as a characteristic of any material item not devised by the mind or produced by the hand of man.

Natural Dialectic is one of different models mankind has used to try and understand the universe into which each one of us is, without asking, born. With roots deep in human thought and variously expressed at different times and places, the Dialectic's 'philosophical machine' reflects an oscillatory framework within which nature operates.

Any philosophical infrastructure, whether mathematical or verbal, is built of symbols, that is, of code or language. Language, involving a specific assignment and coherent arrangement of symbols, is the way that meaning is organised and information conveyed. Therefore, in order to understand, think of or communicate messages, it is helpful to have clearly grasped whatever particular grammar is being used.

In this case, does **Natural Dialectical Philosophy** accurately reflect the *modus operandi* of cosmos? And, as a whole, construe the grammar of polarity? It is, at least, the dynamic formulation within which the narrative of a parent book, Science and the Soul (*SAS*), is expressed. Accordingly, this 'potted grammar' extends many connections to the more elaborate explorations of *SAS*. It also links to Adam and Evolution (*A&E*) and A Mutant Ape? The Origins of Man's Descent (*AMA?*).

Bold, underlines, italics and red script are simply how a lecturer at lectern easily identifies his flow. Stripped of much case-building narrative, such devices help to pinpoint the principles by which nature's language speaks; and the illustrations are from power-point slides. After careful and perhaps exciting inspection you may better judge whether Natural Dialectic's grammar well interprets nature's text and thereby accurately reflects the logic of creation. So jump aboard and ride this streamlined 'thought machine'. Intellectual seat-belt fastened, we shall travel far and fast from here…

Lecture 1:   Introduction

> **In the wisdom of Albert Einstein's words,**
>
> *"... to seek the simplest possible scheme of thought that will bind together the observed facts".*

> **To generate a neutral framework within which data can, from opposing materialistic and holistic viewpoints, be assigned and reasonably assessed;**
>
> *to generate a philosophical routine, a 'dynamic' within which physic and metaphysic are accommodated and may be reconciled.*

This course has two aims. The first involves, in the wisdom of Albert Einstein's words, **"... seeking the simplest possible scheme of thought that will bind together the observed facts"**. Indeed. To develop, by applying the razor of parsimony to all the world's libraries, an Ockham's cosmic philosophy. The idea is to present a simple, neutral framework within which data can, from sometimes opposing scientific and philosophical viewpoints, be assigned and reasonably assessed.

The second, correlated aim is to compare, as if through different angles of a prism, the logic of our material world (as educed by scientific study) with the perspective that snaps into view if a single immaterial element is added. Such metaphysic is very well known and, indeed, forms the *basis* of our current age. As the industrial age was based on the energy of steam, electricity and, finally, nuclear power so ours is based on information. **Information is the immaterial, that is, metaphysical element.**

We'll methodically compare the two perspectives[1] - call them **materialism and holism** - and the logic of interpretations that derive from them *using the simple models* of a balance, that is, a scale along with a couple of others.

Models are non-verbal descriptions. Pics to hang concepts on e.g. atom. To work successfully their underlying 'grammar', that is, mode of operation must be clear. We shall spend time this first lesson exploring the conceptual vehicle or 'cosmic language' which they help illustrate. If we want a fresh perspective we will need to understand this language, its mode of expression.

If correct, such language should generate an abstract, metaphysical machine, tight-knit, well riveted by bolt and counter-bolt, the simplest working model of the universe. But what is such a vehicle called? How does this philosophical mould work? Science has already derived an objective, mathematical and entirely *materialistic* interpretation of events. How are facts poured into our 'dynamic' so that we can add the *immaterialistic* part of comparison?

At this point I should straightaway scotch the notion that any considerations concerning mind, the origins of the universe or life that are not entirely materialistic are somehow 'anti-science'. If I offer a holistic interpretation of this wall, door or the air am I 'anti-scientific'? Am I 'anti' non-conscious matter or energy - the preserve of scientific study? Holism deals with *both* material and immaterial elements but to say it is 'anti' one of them is absurd. This accusation, though widely employed especially by scientific journals, educators in various university faculties, the media and, specifically, species of materialistic creed, is low-grade and needs be deleted. **<u>Holism embraces science within its compass.</u>**

What's it all about?

Who exactly am I?

Where did I come from?

What am I doing here?

Indeed, is there any purpose in life other than the satisfaction of animal needs?

Then, where am I going?

What happens, when the pot cracks, after death?

And then, of course, the questions of morality, law, society and government both political and personal.

The course will **raise many questions and I expect many of us will meet facts and insights we had not previously encountered**. I encourage questions but ask, to avoid fragmentation, that you reserve them for the end of any particular passage. In which case jot them down with any other of your notes. *Might not these notes be useful anyway for revision in tests and the exam from which you graduate in Fresh Perspectives and receive your degree in Natural Dialectical Philosophy?!*

There will, I expect, be questions to which I cannot immediately give full answers and will have to think about before the next lesson.

More importantly, there may occur, according to your particular level of scientific education, **words or concepts** (say, genetics, psychosomatics or lambda 'force') with which you are unfamiliar. We have not the time for an omni-science course, once round the universe in ten lessons, so that I suggest you make, in your own time, copious use of the most readily available sources of information such as Google and Wikipedia. *And, speaking of fastening intellectual seat-belts, we shall travel far and fast but still may slightly overrun the tight schedule. Do not be concerned as we'll cope with this if it occurs.*

Finally, in this brief preface, let's recall that *all* of science and philosophy is concerned to address the most natural questions of a being that finds him or herself as a body on earth, a planet in space, for a brief spell of time. What's it all about? Who exactly am I? Where did I come from? What am I doing here? Indeed, is there any purpose in life other than the satisfaction of animal needs? Then, where am I going? What happens, when the pot cracks, after death? And then, of course, the questions of morality, law, society and government both political and personal. These need answers reasoned within, I shall argue, one of two possible frameworks - materialism with its sub-creeds of humanism and so on; and holism, as old as humanity, including a science of material reality, with its sub-creeds as well. *Heterodox, I wish to detach from all orthodox creeds.*

Arts and science. This nested Contents Box displays both course and cosmos in terms of science and theatre. Observation, understanding and drama. A play is writ, stage is set and, lo, upon the moment, players enter and a drama is enacted. This is the rationale behind the order of our course.

> **Materialism's axiom is that every object and event, including an origin of the universe and the nature of mind, are material alone; a few oblivious kinds of particles and forces compose all things.**
>
> Although the universe appears to work by rules and to have been established in a very particular way, this appearance of order is in fact unplanned. Its invisible framework of regulation must have occurred by chance and, since inception, individual objects and events (phenomena) occur by chance as well.

Now to business, a short but foundational piece concerning Primary Assumptions.[2]

The first step taken on any journey is critical. If, for example, you intend to travel straight from London to Edinburgh and your first step is east or west you will not arrive.

In the case of cosmic world-view the first step is philosophical but also critical. Is your primary assumption, aimed towards full truth and understanding, correctly set or not? Which answer - materialism or holism - does the evidence best brace?

What are the basic axioms and corollaries of this pair?

Materialism's axiom is that every object and event, including an origin of the universe and the nature of mind, are material alone; a few oblivious kinds of particles and forces compose all things. Moreover, cosmos issued out of nothing; therefore, beyond this realm of physics there is only void; and life is an inconsequent coincidence, electric flickers of illusion in a lifeless, dark eternity.

Although the universe appears to work by rules and to have been established in a very particular way, this appearance of order is in fact unplanned. Its invisible framework of regulation must have occurred by chance and, since inception, individual objects and events (called actualities) occur by chance as well.

> Such axiom must apply to life. In this respect the **Primary Corollary of Materialism** states, by the neo-Darwinian theory of evolution, that life forms are the product of the chemical abiogenesis of a first cell; and, following that, by common descent, of a random generator (mutation) acted on by a filter called natural selection. Such evolution is an absolutely mindless, purposeless process. It is, from a materialistic perspective, a fact so that the *PCM* is a fundamental *mantra* of materialism.

Chance is the creator of diversity. Its scientific *aide-de-camp* is probability. **No matter what the odds, the universe and life *must have* appeared by chance.** Order came about by chance. No telling how exactly, just vague imprecation. Nothing is, perchance, impossible; the sole impossibility is that such a story is impossible. This implacably materialistic narrative is rehashed in every textbook, journal and broadcast.

> **Holism's axiom is that realistic comprehension of the world includes *two* primary components - immaterial and material or, as obvious to everyone, mind and matter.**
>
> **A scientific world-view that does not profoundly and completely come to terms with the nature of conscious mind can have no serious pretension of wholeness.**

A caveat. To materialistically presume that what is not material is not natural and, therefore, does not exist is a first order, pseudoscientific error. To 'pretend' it does exist is prejudiciously judged, by some, 'pseudoscience'. If, however, the basic nature of information is immaterial/ metaphysical then is all IT or, design, engineering or your own thinking simply 'pseudoscience'?

Holism's Primary Axiom **is, on the other hand, that realistic comprehension of the world includes** *two* **primary components - immaterial and material or, as commonly perceived, mind and matter.**

But is there really any difference between the two? Isn't consciousness unconscious? Matter mind? Isn't a material brain the same as, or at least the generator of, your mind? Aren't you your body? It is made of cells, cells are made of chemicals, chemicals of atoms and atoms aren't alive. Why then should even a billion billion of them become conscious? If atoms, molecules and cells aren't alive then your body isn't. Alive is not, even by restricted biological definition, the same as lifeless. Organic bodies might be marvellous machines but are they not alive. *So who are you? Are you alive or dead?* **It follows that a scientific world-view that does not profoundly and completely come to terms with the nature of conscious mind can have no serious pretension of wholeness.**

We can go further. You are, of course, alive. You know full well, subjectively, life's consciousness - but is it proven physical? Your body's doubtless physical and you accept a cosmos made of matter; and if body is a special composition made of universal matter may we not holistically suggest that, likewise, human form incorporates its special part of universal mind? To repeat, is not your mind, informant and informed, metaphysical? And, like your body, natural and part of something universal? Using naturalistic methodology (which assumes only material components and answers that exclude any immaterial element) experimental science cannot prove that life's central part, mind with subjective thought and conscious experience, is just a product of non-conscious particles and forces. *Holism, therefore, simply adds immaterial, as a second fundamental ingredient, to material. Or, conversely, it adds material to immaterial.* Since immaterial is not material it adds nothing physical at all. **But hence follows, it is argued, this philosophy's powerful and impregnable validity.**

The second proposition is, because our understanding reasonably reflects it, that existence as a whole *is* **'logical'.**

Holism's logic must apply to bio-logical life. In this respect its **Primary Corollary** states that the origin of irreducible, codified (or highly informed) biological complexity is not an accumulation of 'lucky' accidents constrained by natural law and death. Forms of life are conceptual; they are, like any creation of mind, the product of purpose.

> The **Primary Corollary of Natural Dialectic** states that the origin of irreducible, biological complexity is not an accumulation of 'lucky' accidents constrained by natural law and death.
>
> Forms of life are conceptual; they are, like any creation of mind, the product of purpose. Such assertion is, in the face of materialism, absolute anathema. Yet, if materialism's first axiom is incomplete then every step that follows will lead further from original truth. *An axiom that discounts the force of information may well be largely incomplete.*

Such assertion is, in the face of materialism, absolute anathema. Yet, if materialism's first axiom is incomplete then every step that follows will lead further from original truth. *An axiom that discounts the force of information may well be largely incomplete.*

Furthermore, let us at the outset be completely clear - the basic assertions of both materialism and holism are philosophical; *neither is a scientific one.* Holism includes metaphysic; materialism, *ad hoc*, excludes it. Material science can never prove holism's metaphysic, based on mind and information, is untrue. *In this case, if the holistic axiom that mind and matter are two different kinds of element is true then holistic logic in its entirety is unassailable.* **The previous three slides express the holistic paradigm and preface its way forward.**

Three main points arise.

1. Such axiom exacts a toll. We need answers:[3]

(i)   the nature of consciousness, sub-consciousness and non-consciousness.

(ii)   whether individual mind can exist independent of a body and, if so, the nature of its entry, attachment, exit and disembodied condition.

(iii)   the interactive relationship of individual mind with body; the nature of any *PSI* (psychosomatic border or, perhaps, quantum linkage) between mind and matter.

(iv)   the mechanism by which universal mind, if such exists, might inform non-conscious forces, particles and gross phenomena; the origin of physical constants and patterns of behaviour, that is, the laws of nature.

(v)   the nature of physical and biological prototypes, homologies or, if any, archetypes.

(vi)   the question whether biology is informed by chance and aimless natural law or by design in accordance with such law; a wholesale reappraisal of the neo-Darwinian theory of evolution.

That last one might, as already hinted, raise an eyebrow. After all, a creed has no serious authority unless it can give a satisfactory answer to the question 'where did *I* come from?' Therefore this theory, seized upon by 19th century humanism, is by now a well-packaged web of stories told repeatedly to the masses until it has become a contemporary truth. However, as is usual in science, if a previously unconsidered factor emerges (such as dark matter in physics) then reappraisal of a paradigm has to occur. So, we'll see, with neo-Darwinism.

2. We need to nail down the language (one not exclusively materialistic) in which holistic answers and arguments can be included - **Natural Dialectic.**

Dialectic (or dialectical method) is to-fro discourse between opposing views in order to establish reasonable truth. Of course, there may exist several antagonists in a debate but, in the case of nature, it may be shown that only polarity or, at most, trinity, holds sway.

Natural, for its part, is often construed as a characteristic of any material item or construction not devised by the mind or produced by the hand of man.

Any philosophical infrastructure, whether mathematical or verbal, is built of symbols, that is, of code or language. Language, involving a specific assignment and coherent arrangement of symbols, is the way that meaning is organised and information conveyed. Therefore, in order to understand, think of or communicate messages, it is helpful to have clearly grasped whatever particular grammar is being used.

In this case the dialectical framework represents, essentially, cosmic infrastructure and is thus, simultaneously, another formulation of mankind's Perennial Philosophy. Simply, it asserts that to-fro, binary logic is the way that all things work; and is, furthermore, the way our polar cosmos is constructed. Does this assertion accurately reflect the *modus operandi* of cosmos? Does its streamlined 'thought machine' well construe the grammar of polarity? We shall explore this after checking the third point.

3. **Universal models of creation.**[4]

a) You can think of cosmos as a pair of scales. Balance, up and down.

Such scale integrates a pivotal, balancing factor (*Essence*) with two antagonistic vectors of *existence*. In this, the perpetual changes of creation are seen as myriad adjustments against the disturbance of balance. Existence is a field of ceaseless change. *Creation's wheel is, therefore, one of relative uncertainty and instability. **Such* eccentric instability is forever perfectly imperfect; perfect imperfection shows as the perpetual motion of continual changes at local times and spaces.** Existence thus amounts to a swinging but self-regulating balance; it amounts to the sum of myriad individual actions and reactions each of which always shows proportions of two vectors pivoted around the poise of a third non-vector.

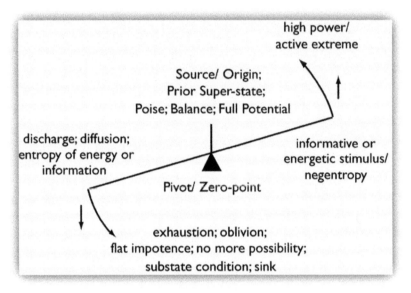

Now inspect this triplex 'stack':

↓	down/ descent	Point of Balance	up/ ascent	↑
	sink	Source	flow	
	negative	Neutral	positive	

Note, for future reference, its trinity, the vectors (up and down); capital letters for the central column; and flow (or relative action) between Source and (relatively inactive) sink.

b) You can also think of cosmos in *dynamic* terms of <u>concentric rings.</u>

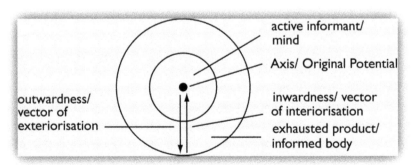

12

This figure of concentric rings includes antagonistic vectors to and from its Central Source. Is this not, from generative seed to petals, how the patterns of our cosmos all concentrically flower?

We are building a picture of spectrum, that is, scale of energy or information. Natural Dialectic calls it the conscio-material gradient, or spectrum, of creation.

c) Thirdly there is the idea of <u>Mount Universe, the Cosmic Pyramid.</u>

A cone or, squarer, pyramid describes '*static*' hierarchy. A useful representation is the stepped pyramid, also called a *ziggurat*. In this case each step of a **ziggurat** stands clearly for a phase, level or stage; and the apex, its capstone, a point that points beyond the finite grades below, implies peak infinity. **This capstone is the highest point and source of what we call Mount Universe**.

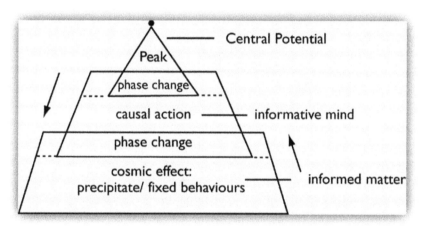

This slide illustrates some triplex stacks describing a dialectical ziggurat.

If up and down (or motion) marks the basis of existence we see **duality**. Change and relativity are the nature of all things. But it is obvious, from the triplex stacks, that we are also dealing in **trinities**. The nature of this Central, Balancing Source is of interest.

Ups-and-downs and peaceful poise; ins-and-outs and equilibria; existence is an ever-changing play, a field of pulsing tensions that evinces, as already mentioned, trinity. *If the triplex is naturally fundamental it is also fundamental to the operation of this Dialectic; but while the orient has long worked and woven with these immaterial radicals, these fundamental threads, then western minds have not.*

Take a scale's three phases - balanced and swinging upward or downward. **The actions that these words symbolise find various expression in each object and event.** More subtly, they reflect the

| ↓ | down/ descent | Point of Balance | up/ ascent | ↑ |
| | sink | Source | action/ flow | |

qualities intrinsic to creation. They compose three fundamental operating principles whose permutations are expressed, in varying degree, as *tendencies* both psychological and physical. We call them cosmic fundamentals.[5]

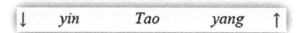

| ↓ | *yin* | *Tao* | *yang* | ↑ |

You might be familiar with this far-eastern, Taoist triplet and its associations. Therefore, to head its stacks, Natural Dialectic (ND) uses an equivalent, equally ancient abstraction whose purity is, because it is less familiar, less stained by prejudice or shadowed by prior connotations. Let's employ (despite, perhaps, conservative resistance) the following as-yet faceless trio.

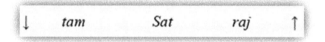

| ↓ | *tam* | *Sat* | *raj* | ↑ |

Sat (equilibrium), *raj* (action ↑) and *tam* (inertialising or materialising tendency ↓). Those interested in dietary cooking might have already picked up on their categories. For example, *sat* food is fresh, light and includes fruit, vegetable and cereal products. *Raj* food stimulates; it is hot, promotes physical activity and includes spices, curries etc. A *tam* ingredient is stale or heavy; it includes meat and alcohol. You get the flavour. **These fundamentals are all-pervasive to the extent, as we shall see, of describing the three major tiers of Mount Universe itself.**

We're almost there. The final step is to treat these triplex stacks as dualities.

negative	*positive*
fall	*rise*
exhaustion	*stimulus*

Let's start with a simpler 'duplex' presentation of the case. **This is because Dialectical stacks are a columnar expression of polarity.**[6] A *stack* is a set (or pile) of members. Each *member* of a stack (e.g. negative *and* positive) consists of a pair of polar 'anchor-points'; and each *element* of the pair (negative *or* positive) is arranged according its

14

fundamental characteristic viz. its tendency to (*raj* ↑) rise or (↓ *tam*) fall. Health/illness, life/death, happy/ sad - the list is endless.

It is immediately obvious that members of a stack are not synonymous. *But they equate in fundamental character; they represent equivalence in terms of cosmic fundamentals.*

Also, each member implies a scale or **dynamic range** that runs between its elements, that is, its '**paired opposition**' or '**complementary covalency**'. For example, you can have 'more or less negative'; you may suffer a 'greater or lesser' fall. *As a scale of greys runs between extreme black and white, so it is implied that a scale permits oscillation of values between any pair of elements.*

On any scale motion runs down (↓) and (↑) *vice versa*. Or you could visualise a horizontal spectrum with your arrows oscillating (←) down or (→) up. For our example: light and dark.

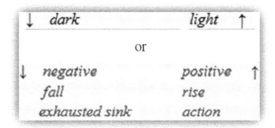

Of vertical and horizontal visualisations one is superfluous. Edison decided, with the flip of a coin, to designate electron – and thereby proton +. Dialectical convention prefers the vertical representation because this represents the rising and falling qualities of cosmic fundamentals and thus universal activity. **A single, non-repetitious first-line representation elegantly indicates the direction of elemental vectors for the whole stack.**

What about the central (*Sat*) character? Sometimes more convenient to write **a non-vectored di-logical stack and then polarise its neutrality into a second (*raj*/ *tam*) vectored and also di-logical membership. What is meant by this? How is it done?**

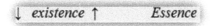

The first, top stack of the pair disposes Essence and Essential Characteristics on the right against those of existence on the left.

polarity	*Neutrality*
duality	*Unity*
motion	*Inaction*
existence	*Essence*

Such a binary stack is called Primary, Essential or Central Dialectic.
It is indicated by writing the right-hand column with a capital letter.

A Primary Stack sets (*Sat*) Unity against (↓ *tam/ raj* ↑) duality; or, if you like, it sets qualities of motion against Inaction or relativity against Absolution.

↓ *tam*	*raj* ↑
negative	*positive*
divisive/ polarising	*depolarising/ unifying*
exhaustion	*flux*
lock-up	*freedom*
materialisation	*dematerialisation*
creation	*dissolution*

Primary, Essential or Central Dialectic.

tam/ raj	*Sat*
existence	*Essence*
polarity	*Neutrality*
expression	*Potential*
limitation	*Infinity*
duality	*Unity*
relativity	*Absolution*
motion	*Balance*
something	*Nothing*
(3)/ (2)	*(1)*

secondary existential or polar dialectic:

↓ *tam*	*raj* ↑
fall/ down	*rise/ up*
negative	*positive*
division/ multiplication	*unification*
isolation	*connection*
drag	*stimulus*
(3)	*(2)*

Duality, however, implies polarity. **Such polar component is expressed in the lower, vectored so-called** <u>secondary, existential or peripheral/ polar dialectic</u>. Thus **secondary, existential stacks** (*written exclusively in lower case*) represent the various kinds of polarity from which the changeful web of existence is composed.

Here I'd like to elaborate, very briefly, on three characters from the right-hand column of the previous Essential stack - Infinity, Zero and Unity.

finite	*Infinite*
1/ on/ action	*0/ Off/ Peace*
object/ event	*Nothing*
polarity	*Neutrality*
duality	*Unity*
↓ post-active effect	*causal stimulus ↑*
no-change/ apparent fixity	*changes*
switched off	*switched on*
unit/ object	*flux/ event*
unreactive/ neutral	*charge-based reaction*
peace/ zero action	*action*

These are important 'leaders' in the way cosmos is projected from its Essence.

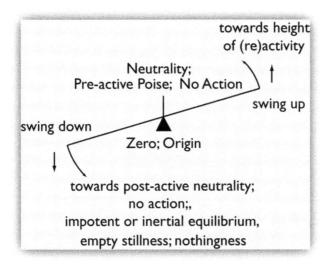

No doubt, the characters of the Essential column are commonly *reflected* as part of creation *but only in local, temporary forms*. Every existential object or event has boundaries; these bestow psychological (in the subjective case of mind) or physical (in the objective case of bodies)

17

condition. Thus we call such realities *'lesser'*, *'seeming'* or *'apparent'*. They are *phenomenal* and finite in some degree. Time and space are interesting *lesser infinities*. Other phenomena (including balances, potentials and so on) are relatively restricted, that is, localised in time and space.

issue	*Source*
motion	*Inaction*
↓ *inaction*	*action* ↑
sink/ immobility	*mobility*

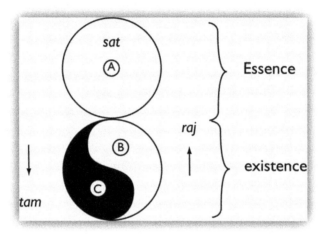

tam/ raj	*Sat*
below	*Transcendence*
range of action	*Super-State*
expression	*Pre-motive Potential*
current	*Source*
↓ *tam*	*raj* ↑
descent	*ascent*
subtending	*transcending*
deactivation/ fixation	*rarefication*
subtendent base/ sink	*stimulus*

These three slides illustrate a phenomenon I'd also like to touch on called reflective asymmetry or inversion.[7] As regards cosmos (and, therefore, your microcosmic self) such inversions are from the expression of Informative Potential to its complete absence (or locked,

18

impotent fixity) in non-conscious, automatic matter; and, as regards energy, a fall from subtle to gross, locked expression of solid matter. These are the couple of *Primary Reflective Inversions* that, interlocked, compose the **cosmic** conscio-material gradient of creation or, if you like, the informative/energetic scale of cosmos.

Now let's briefly check how stack and model integrate.

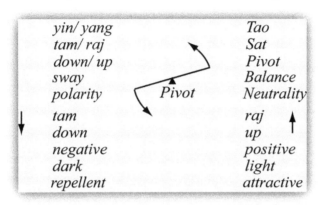

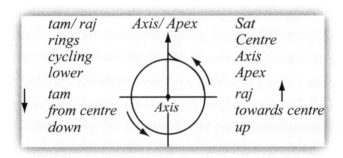

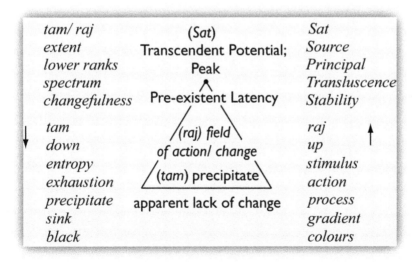

19

We are now in a position to ask, "How is all this related to science"?[8] How can stacks and cosmic fundamentals re-angle not the facts but the prism of our perspective?

> *A simple link from <u>psychology</u> describes the three basic conditions of information as (sat) potential informant (concentrate of consciousness) prior to (raj) active mind ((both informant and informed) and (tam) passive, dormant or sub-conscious mind.*
>
> *A simple link from <u>physics</u> describes the three basic conditions of energy as (sat) potential prior to (raj) action and (tam) exhaustion.*
>
> *Thirdly, a simple link from <u>biology</u> describes the three basic conditions of life on earth as (sat) informative archetype, (raj) metabolism including a code carrier called DNA and (tam) finished product called developing or adult body.*

We can now see and will elaborate the triplex logic within which we might profitably and comprehensively treat the sciences of (mind) psychology, matter (physics) and their connection, mind-with-matter, in biology. <u>What insights await?</u>

We now understand that there exists a holistic, triplex (threefold) treatment of science as well as the one-tier materialistic one; and that, as regards holistic logic's fundamental connectivity, there are thousands of possible stacks. I have used just aan exemplary handful in this lecture. You can make them up yourself. *What they amount to is a system for generating fresh perspective, links you have not made before, new insights into cosmic connections.*

To summarise: **Natural Dialectic asserts that to-fro, binary logic is the way that all things work; and is, furthermore, the way our polar cosmos is constructed. Such radical infrastructure involves a trinity of qualities.**

Henceforward we'll see how the fundamentals fit with science, building Mount universe and, next lecture, we'll investigate the question of immaterial information.

Lecture 2:   Information

We draw a distinction, elaborated in the section on psychology, between **subjective focus**, contemplation which allows perception of principles; and **outward focus** by means of sense, manipulations and their technological extensions. These 2 modes of information gathering and processing will be discussed later in the section on psychology. For now simply note that these ***anti-parallel vectors*** of mind[9] correspond to what we call *bottom-up* and *top-down* directions.

Bottom-up is, broadly, that of the person working from detail to understanding patterns and principles by experience. This kind of logic, **from detail to principle, is called inductive**. It is the way of naturalistic curiosity, that is, of experimental science.

Top-down is, broadly, that of the expert working **from principle and its application for making sense of details.** This kind of logic, **from principle to detail, is called deductive**. It is the way, from mind to matter, that holism works.

There is a third, important sort of logic that we'll come across. **It applies to unique historical events or futuristic speculation that we have neither seen nor can for sure repeat.** It is called **abductive or conclusion by best inference.**

The Flammarion shows a medieval scholar seeking to rise from earth to the abstract heavens!

Cambridge University itself was founded to **e-ducate** from lower mind (or the reflex, animal condition) and promote the *intellectus* or *higher mind* in which **principle, a form of highly-condensed and**

21

connective information, is better understood. It is the way, with better panoramic view, of the commander on a hill. The higher the point of reference the better!

Did not the student, sitting on a tripod (three-legged stool) achieve his degree by successful to-fro, dialectical discussion with his tutors? In dialectical respect I'd like to draw attention to duality within unity and the basic elements of the existential dipole we have drawn.[10] We will shortly relate its triplex nature to Mount Universe.

existence	Essence
issue	Source
lower	upper ↑
non-conscious	conscious
informed	informant
reflex	creative
matter	mind

This stack describes the duality within unity of the diagram below.

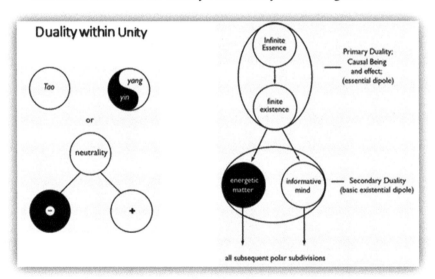

Materialism has decided matter is the basis of creation (though it does not specify the nature of the nothingness that matter must have started from); thence mind is an excretion from a brain and soul a pure (but physical) imagination! Holism's revolution turns materialism on its head. The diagram below shows The Monopole, the nature of Sole Cosmic Independence - Infinite Essence; and its dependent issue, all the motions of existence. We are now in a position to relate cosmos in terms of **the triplex yin-yang-*Tao* model and the cosmic pyramid: (*Sat*) Essence with (*raj*) mind and (*tam*) matter.**

Upper Pole - Information
(Sat) Potential Information
(Raj) Active Information
(Tam) Passive Information

Lower Pole - Energy
(Sat) Potential Energy
(Raj) Active Energy
(Tam) Passive Energy

This small table is amplified in the models of Mount Universe shown below.[11]

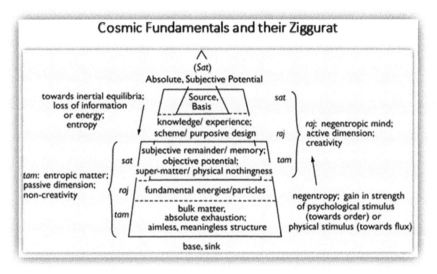

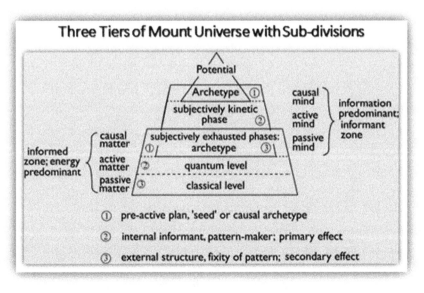

23

We'll meet the upper and lower sectors of this ziggurat in Chapters 3 and 4 respectively. Chapter 5, biology, inhabits both.

The place of information in this scheme appears below. It will be further elaborated in Chapter 4.

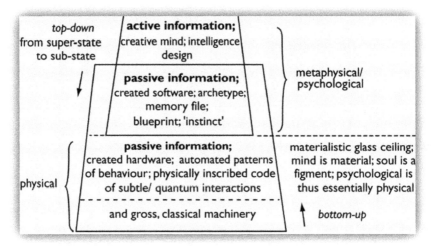

Following the nature of these three cosmic fundamentals every event (including creation itself) involves prior precondition (called potential), an active phase and a passive, finished or fixed phase.

A slide illustrates this division for the treatment of information and energy. So we've looked at divisions of Mount Universe, not least with respect to information.

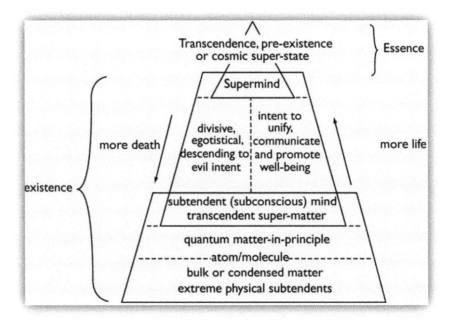

Before we elaborate on what we mean by 'information' I would like to take a brief look at the Mount's extremities, that is, its extreme initiation thesis and consequent antithesis. This is because the Transcendent Super-state is regarded as the Centre, Source or Pivot round which cosmos swings. **Why?**

On the other hand, materialism asks why eternal matter of some species shouldn't be the source of all there is. **We'll deal with why science thinks all physical phenomena were caused, material cosmos had a start and steady-state theories of eternal matter have been binned in the section on physics.**

Why should Super-state hierarchically precede mind as well as well as matter? We'll check that out both later today and in the section on Psychology.

It is true, *top-down*, that products subtend their producers. A producer stimulates and stirs things up. This producer may be mined or, materially alone, a stimulant energy (such as, for example, Chladni's dish in Lecture 3, whose musical vibrations create organic shapes in sand). *From this perspective active informant precedes its passive, informed consequence, metaphysical mind plans physical arrangement and super-natural precedes natural order.* **Mind first, body after; body's an appendage of its mind.** It happens all the time with us: idea (or desire) precedes outcome. *Does it, that physical depends on metaphysical, invert your 'normal', that is, 'sensible' sense of things?*

So, for example, we say that sleep subtends waking, sub-conscious subtends conscious. And, conversely, waking transcends sleep. It is like the various bands of a spectrum UV, visible (us) to IR and black. We're emphasising a notion of **hierarchy** central to a tiered universe. **Spectrum** introduces an idea of hierarchy, scale or, in ancient parlance, Jacob's ladder.

Now the hierarchical interest turns *to causes*. Of course, we can think of causes in terms of knock-on effects. They impact, in physics, chemistry and psychological reflexes, in '*horizontal*' chains. **No doubt, energetic causes push effects; they bump you from behind; their arrow, physic's arrow, runs from <u>past to present</u>.** And things suffer (though you'd hardly think it as regards a proton or electron) from increasing weariness called entropy. They run out of steam. We call this <u>horizontal causation</u>.

What, though, about a cause that is conceptually implanted? What about not energetic but *informative causation*? **This is goal-oriented; and goals are <u>in the future pulling you their way</u>. They pull you from the future; they lift forward. They are metaphysical *attractors*, guides that govern your behaviour as they lead you through the world.** Not material but immaterial, such leadership is not by force of gravity, electric charge or magnetism; information is metaphysical not physical; it's psychological and, although at this point you may want to know exactly how, be patient - as the course elaborates we shall come to see. *Information's entropy is negative.* **Mind is negentropic and thus metaphysic's arrow flies, from future back to present, anti-parallel to physic's.** Thus you are guided, present to the future, by plans or programs that realise their goals. Example, complex rules of a game of football.

Waggle your little finger. What *caused* what? The line of operational command runs from conscious, informant mind through sub-conscious templates (memories, instincts and so on, dealt with later in Lecture 3) over a psychosomatic border to innervated body. Between the zones of purpose and non-purpose, incoming or outgoing information would be first translated to, or last translated from, the physical side by the product of 'excited' electrons, that is, electromagnetism. Where *matter subtends mind*, this 'radiant' phase would in turn be subtended by the biochemical; and both levels occur within bulk, biological structures, that is, cells, nervous system, muscles and a coherent whole body. **This places your finger-waggle at the base of an informative/ energetic hierarchy.**

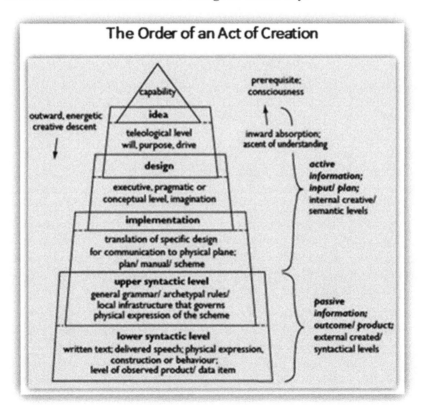

This process we call _vertical causation_. **It is the order of an act of creation.** It drops from creative mind to created material form. As a later talk will elaborate, your own sensible, physical form (perhaps including the origin of its shape) resides at the bottom of such a hierarchy. Bio-form's a frozen yet dynamic program; or, prosaically, a 'functional structure'. Very complex, yes; automated, yes; body's an incredible machine.

In this respect holistic view holds cause in some cases _higher_ than effect.

The finger-waggle indicates how you personally operate. However, we can reach beyond the normal conscious mind. We now focus on Source, Centre or Highest, All-Informant First Cause.

Aristotle believed in a First Mover itself unmoved by any cause. St. Augustine observed that no 'efficient cause' can cause and thus precede itself. Thus causal order can't be infinite; there needs to be an uncaused primal cause. Existence is composed of caused, finite events. Whatever begins to exist, asserted the _sufi_ Algazel, has a cause; _and 'something which begins has a sufficient cause' is also the modern principle of causality._ **This principle is constantly verified and never falsified. The physical universe began to exist and therefore has a cause.** What is caused is not eternal. It is finite. Its effect becomes a further cause. Thus all existence is a changeful network made of causes and effects; creation is an action and reaction zone.

Where nothing is an absence, nothing comes from nothing. As regards the origin of physic, did you think space was nothing? Wrong. Vacuum did not come before its cosmos; nor, since particles arise from fluctuations in it, is it nothing. Plenitude of vacuum is a new-found quantum fact. Why, therefore, should cosmos as a whole derive from nothing? If it didn't, did the universe create itself? Then it existed prior to itself. Cause caused itself. Such is the kind of incoherent, 'boot-strap' logic some cosmologists devise. Perhaps, you claim, there was no cause of physicality! But, if there were, such cause could not be physical. It must transcend the physical. Its physical non-being must be immaterial; its being must be metaphysical. This Uncaused Being is not absent. It is self-sufficient, potent and with presence; its causal level of reality is 'higher' than non-conscious, physical phenomena.

In summary, it's as simple as it's crystal clear. **What starts to exist is always caused.** We presume the physical universe started and eternal matter, that is, endless change, is not a feasibility. Therefore, material cosmos, known as nature to the natural sciences, started to exist and has a Natural Cause. _This cause, preceding physicality, is physical non-being._ It is nothing physical but causes cosmological effects. What came (or comes) before the latter's matter, space and time must itself be time-less, space-less, immaterial. Preceding its secondary causes and effects the primary, first cause must be physically uncaused; it is uncreated in a naturalistic sense, thus super-natural. Metaphysical.

Rephrased, motion is potential that's expressed; cosmic action is Uncaused Potential worked. The primal motion of Essence is First Cause; *such First Expression is the Start of starts and purest form creation issues in.* All existence, psychological and physical, is the Start's effect. This is to say, for example, **the term for lower, physical First Cause preferred by Natural Dialectic to 'big bang' is '_transcendent projection_'.** The latest version of big bang may or may not reflect the physical part of truth but, for certain, it was metaphysically conceived.

What, therefore, is the Nature of this Informant Metaphysic?

The first thing to note is that, in cosmic terms, *physical first cause is* **not** *the same* as *Psychological First Cause* (Pre-motive, Essential Super-state). In short, as we'll see in Lectures 3 to 5, there are two kinds of first cause, the **higher conscious** and the **lower unconscious.**

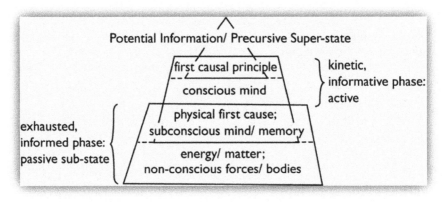

Primary (Metaphysical) and Secondary (Physical) First Causes

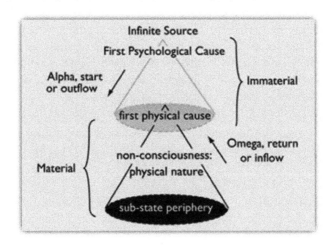

28

tam/ raj	*Sat*
below	*Transcendence*
range of action	*Super-State*
expression	*Pre-motive Potential*
current	*Source*
↓ *tam*	*raj* ↑
descent	*ascent*
subtending	*transcending*
subtendent base	*action*

Now, after this preamble, it is time to focus on the nature of Natural Dialectic's immaterial element, **information**.[12] We are information junkies. We want the news, crave more information and want to know everything! Perhaps not only externally in the universe (including our bodies) but on the immaterial, contemplative mind side too. So what is info? It is defined as facts or fictions learned, knowledge, understanding or intelligence (e.g. sigint); also as computer data or what is conveyed by a particular arrangement of things or symbols. The arrangement of symbols, representations or stand-fors is called code. Language. **But is there any kind of information without mind?**

Top-down, *to receive and signal information is the immaterial gift of mind and never matter.* **It is irreducible to scientific scrutiny. Yet, as we'll see, its 'semiotic' metaphysic dominates debate about creation and our lives.** Pick up a postcard, menu, letter - anything informing you. The object is reducible to chemistry and physics but the *sign* or *message* it conveys is not. Signs and signals always have a purpose; objects *per se* never do. In other words, information is fundamental to our being but does not fall, in a semantic sense, within the scientific remit. (cf. telephone blips) *Particularly, mindless origin of bio-code is an irrational hypothesis.*

Bottom-up, **however, everything is seen as energetic interactions. Energy's the physical informant.** Information's therefore, even in the case of brain, evolved by chance and natural law. **Oblivious physic generates its off-chance but never has a goal.** *Particularly, mindless origin of bio-code is, from this perspective, a rational hypothesis.* Is such 'metaphysical consideration' an illusion, a confusion or the truth?

Firstly, in this dilemma of perspective, let's distinguish mundane (normal, meaningful) sense from scientific/ materialistic understanding. *In mathematics of the latter sense 'information' must not be confused with 'meaning'.* So what exactly, in this apparent shortfall, *does* the word mean?

29

It was Claude Shannon who, with Warren Weaver, first devised a mathematical and thereby scientifically acceptable definition of information. Shannon treated its transmission in purely physical terms according to statistical formulation of the entropy-inclusive laws of thermodynamics (2nd Law is perhaps most basic and tested in physics, no exception ever observed). In such transactions his unit was the *bi*nary digi*t*. This on/ off, one-zero 'bit' allowed the quantitative properties of strings of symbols to be formulated. His theory inversely relates information and uncertainty. The more uncertain, the less probable a sequence of symbols or arrangement of materials is, the more information it is calculated to contain. Rephrased, an amount of information is inversely proportional to the probability of its occurrence by chance. Simple, repetitive or predictable sequences contain less and complex, irregular arrangements more information. Thus 'Shannon information' is a simply measure of improbability. *It makes no judgement whether such irregularity is specified; it involves no sense of meaning.*

ZNQW&RSIXT AZ2HVB
COMPREHENSIBILITY

Shannon's definition is suitable for describing statistical aspects of information such as quantitative aspects of language that depend on frequencies (such as how many times the letter 'a' or the word 'and' occurs) but it treats any random sequence of symbols as information without regard to its concept, meaning or purpose. In other words, the more improbable any arrangement the richer is its Shannon-defined information content. *In short, two messages, one meaningful and the other nonsense, can be exactly equivalent according to this form of analysis.* For example, ZNQW&RSIXT AZ2HVB and COMPREHENSIBILITY are assigned the same value. In other words, Shannon has reduced information to a statistical quantity; he has shaved off any sense of meaning, cut out sense of purpose in numerical analysis. *Shannon-shaving deals in* <u>*non-purposive complexity.*</u>

30

In this mode of thinking a 20-letter randomly-generated and meaningless sequence contains information richer than the simple phrase 'I love you'.

It is important to grasp how Shannon, thoroughly negating logical reality, accords randomness and purpose equal status. **But randomness is really reason-in-reverse; it is information's opposite.** So when it comes to language or to mechanism, including those embodying biology, such conflation is an error of first order. Shannon's analysis does not distinguish between presence of mind, authorship or creativity and their absence; it fails to recognise purposive specificity; nor does it accord function or meaning any premium. *Thus order and precise meaning - usually complex, always accurately coded - might as well be able to emerge from randomness or senseless motion under natural laws of physics.* You can thus deceive yourself and think in topsy-turvy terms of 'senseless design', an anti-teleology well-known as Darwinian evolution. Such statistically-generated self-deception works well because, like an ups-a-daisy religion, its adherents believe it. *It infects, for sure, the mind-set of an academic discipline throughout the world. This is why the issue is a serious problem. We are being fed, as a matter of faith, nonsense about ourselves.*

Informative capacity inhabits the *arrangement* of material - not just any old arrangement but one with meaning that serves purpose or accords with principle. For example, everything around you in your room, not least the simple chair you are sitting on, involves informative capacity. Purpose weighs no more than understanding; both weigh just as much as meaning. In the scientific balance meaning weighs as much as abstract theory. Each is lighter than a feather. Universal mind would weigh precisely nothing yet remain a vital hidden variable. *If, though, information is absent massively, what is it?*

Norbert Wiener, mathematician and founder of cybernetics and information theory, said *"Information is information, neither matter nor energy. Any materialism which disregards this will not survive one day."*

Information is not a property of matter. Purely material processes, unguided except by natural law, are fundamentally precluded as *sources* of information. That is to say, there is known neither law nor process nor sequence of events through which oblivious matter can create or collect information.

If immaterial then, as this volume shows, it renders materialism with scientism (but not science) most illogical! Indeed, Norbert Wiener, mathematician and founder of cybernetics and information theory, said *"Information is information, neither matter nor energy. Any materialism which disregards this will not survive one day."*

There is, furthermore, known neither law nor process nor sequence of events through which oblivious matter can create or collect information. The latter is not a property of matter. Purely material processes, unguided except by natural law, are fundamentally precluded as *sources* of information. Information is not a thing itself but a representation of physical things and metaphysical entities. Physical nature wears no numbers, bodies are devoid of words; these are planted or taken away by mind.

To rephrase, information involves the transformations and arrangement of material phenomena but such phenomena *per se* are purposeless; they can inform a mind (by sense, mathematical descriptions etc.) but not their own aimless, mindless, non-conscious oblivion. However, although the default materialistic position is to claim laws of nature operate within a field of chance encounters, it is possible to holistically argue that physic provides a dynamic stage for purpose, that is, for life. *There may be a purposive source (or program) that informs specific, automatic behaviours.*

In this case, let's take a second angle on the way that cosmos and the human individual are informed.

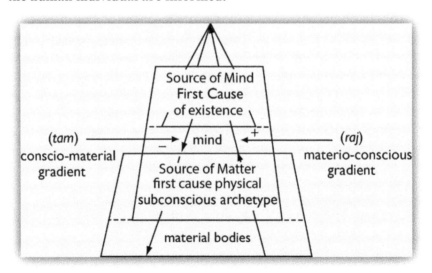

This is a key illustration. Its ziggurat reflects the previous diagram 'Two Kinds of First Cause'; it illustrates the anti-parallels of a dynamic conscio-material gradient that run up-in or down-out. This drops from

Clear Formlessness at peak through psychological and physical forms of various quality. The latter are trapped in reflex, oblivious cycles; the former involve, at least for humans, a measure of choice.

For **'inside information'** you can see the way to go; and how, conversely, sensation leads you 'down' and 'out' into the world **'outside information'**, that is, of physical events. You can thereby easily understand why the sage turns his contemplative focus inwards towards unification with First Cause; and a scientist outwards to the details and, in principle, unification of an essentially non-conscious universe.

Source, before even the primal motion of First Cause, <u>is</u> a concentrate or super-state of consciousness. Such purity is the Uncaused Cause. And consciousness is, by nature negentropic and thus its 'creation express' develops orderly. It devolves down to oblivious physic within whose context chance may play a part.

In other words, inmost centre is expressed outward. Seed is related to every part of its body. Top-down, non-conscious, sub-state physic derives from metaphysic. In this respect metaphysic (higher up the ziggurat) is seen as *within* physic; physic is expressed from it. Paradoxically, as e-m waves translate into sound from a radio but are all around it too, so metaphysic *contains* the automatic, physical universe within itself as well. *After all, why should metaphysic, in the psychological form of universal mind, obey the laws of physic?*

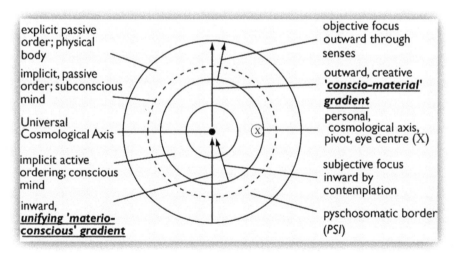

Where peak is central axis this diagram is another way to understand the previous key illustration. *And if we refer back to an illustration called The Order of an Act of Creation we can combine all three models to demonstrate <u>the triplex nature of information according to the fundamentals sat, raj and tam.</u>*[13]

↓ *outward*	*inward* ↑
outside/ outer	*inside/ inner*
down and out	*up and in*
objective	*subjective*
physical focus	*psychological focus*
diffusive/ centrifugal	*centripetal/ concentrative*
reflex/ involuntary	*deliberate/ voluntary*
sensation	*contemplation*
material side	*immaterial side*
materialistic exclusivity	*holistic inclusivity*
atheistic tendency	*theistic tendency*

relative illusion	*Truth*
less right	*Right*
critical comparison	*Criterion*
lower qualities/ lesser values	*Quality/ Value*
shades	*Illumination*
degrees of incomprehension	*Clarity/ Clear Mind*
lower principalities	*Principal*
death/ life	*Pure Life*
lesser truth	*greater truth* ↑
blinder mind	*clearer mind*
away from 'right mind'	*towards 'right mind'*
lower qualities/ separators	*higher qualities/ unifiers*
quantifiable objects/ events	*experience*
non-conscious forms (matter)	*conscious forms (mind)*
individual/ local contexts	*principles/ symmetries*
less/ least real	*more real*
towards darkness/ oblivion	*towards light/ understanding*

First in this triplex issue comes (*Sat*) **Potential Information** from the cosmological axis or source of mind. Such conscious, inside information is what the contemplative sage is fundamentally after. Look *top right* where the Buddha's hand is pointing. It leads up towards Absolute, as opposed to relative, Truth.

And, in the opposite direction down from that Truth, will-power commands and a command's an order. A command initiates consequent action. Authorisation is causal. It says 'Arrange things as I want'. Such is the teleological order of *Kalma* or *Hukm* (Command, as the Muslims call it), *Shabda* (Sound or *Bani*, as the Sikhs say), Zoroastrian *Sraosha*, Taoist *Tao* , Jewish *Ain Sof* or Hindu *Nad* and *Om*. (Note the connection with **sound or vibration**).

Om is, in Christian terminology, Amen. Or, again in Christian terms, Byzantine *Christ Pancrator* is identified as *Logos* (the Word of Reason). Intent in nature; nature's teleology. Thus, of course, from *Pancrator* ('all-powerful') to *Pancreator* is no step; power psychological includes the attribute of creativity. *Logos* is a Greek noun meaning 'reason' or 'the words expressing it'. Speech, message, code. And also 'voice'. *A word has meaning and its purpose is communication. Information. Instruction. Command.* Call *Logos* The Informant.

This was the ancestral view. Ubiquitous but now, in a tsunami of materialism, dipped out of sight. Holistic logic, by its power, demands it will return.

(*Raj*) Active Information is only created in conscious, choice-flexible mind. *What constitutes an act of creation?*

tam/ raj	*Sat*
expression	*Idea/ Purpose*
↓ *tam*	*raj* ↑
passive element	*active element*
informed	*informant*
physical elaboration	*conceptual elaboration*
created outcome	*creativity*
output	*input*
end-product	*development inc. manufacture*
hardware	*software*
machine	*program*

Idea is the seed. There accompanies intention to elaborate the embryonic inspiration's promise. This, grand or trivial in scope, is the basis of all purposes. Can we further understand the order of an act of creation from the way we seek to elaborate, that is, express an idea? From trivial creation of finger-wagging up to creation of a work of art or science all involves, essentially, the same order of process or, if you like, **phased intent**.

The outcome may be simple (as in finger-wagging) or complex. So **complexity** is of **two** kinds:

Non-purposive, natural or entirely energetic (e.g. universe not make a cup of tea in a billion years) made passively derived aggregations including, as in the case of, say, snowflakes. Such complexity is mindless, aimless, passive, purposeless.

Purposive complexity works the other way. *It is a function of information gain, expansion of* consciousness *and a capacity to grasp principles and purposely, creatively exploit the principles and possibilities inherent in any circumstance.* An increasingly concentrated focus of attention wakens to greater capacity, flexibility and possibilities for specifically ordered, coherent or *active complexity*. At each level of ascent the degree of coherence may improve; the degree of ingenuity, adaptability and innovative complexity may increase. For example, human mind and its society constantly respond, adapt and build new, elaborate structures to combat the age-old exigencies of a problematic life on earth. *Active, purposeful complexity* works against the 'downward' wear and tear of time and chance; it codifies and specifies design - which chance cannot. It is an instrument of biological survival, intellectual enquiry, technology and artistic creation. We continually experience it. Its proof, in artefacts and actions, pervades our lives.

Conscious creativity is filled by information, will-power and desire. *Indeed, will and desire are the psychological equivalent of physical electromagnetic radiation.* Will is like electricity, desire like magnetism. Volitio-attractive; pushy will and pulling 'I want' or desire.

We have what we call 'free will' but, really, it is contextually conditioned, limited free will. Nonetheless, mind's dynamo streams from focus of attention, a concentration of interest; and interest may intend to order things or events according to its own purposes. **Call this executive design.** Pragmatics is set at the level of executive or systems analyst. *Transmission of purpose through the pragmatic level involves, in the outward direction, imagining and planning its realisation; and, on the inward, deciphering another's plan.* This is active teleology. In fact, whenever an intention is physically expressed it works through mechanisms and machines. Machines, which include biological bodies, operate according to physical law and fulfil their function using, in one form or another, energy. They specifically accord with plan and inform the world in ways unguided nature can't. As such they obviously link information (that is metaphysical) with energy (that's not).

Causal reason generates meaning. *All active information is meaningful.* **Implementation** of a plan involves *semantics whose specific meaning transcends the generality of grammar and syntax.* The latter are simply vehicles of reasonable expression; their immaterial symbols are needed to make connection with the material world and thereby order it. They translate mind to matter. Coding and decoding, using speech-through-air, written word or other forms of signal, are core semantic business.

They make sense. Although *meaning* **is the important, active ingredient behind codes, data transmission and storage**, nevertheless these passive instruments of communication are important. **We need frameworks (hardware) within which to manage information (software).**

A command erases randomness; it is a deliberate restriction imposed to cause a non-random outcome. It employs an agent of restriction - sign, symbol, code of one sort or another.[14] *This is the world of signals, semantics. What is spoken is not a matter of chance.* **Whatever is encoded is intentional.** Code and incoherent chance are chalk and cheese. They never mix. Language, other symbols and the construction or decipherment of meaningful communication (called semantics) make up the third level of information. In other words, this is the level at which ideas are framed in code or blueprint before their presentation in material form (*important when considering biology, whose basis - metaphysical basis - is code*).

(*Tam*) **Passive information is the expression, external to conscious, flexible mind, of active information.** Such information may be dynamically exchanged (as in the case of speech or body language). Otherwise its impressions are stored either in subconscious mind (featuring relatively inflexible memory, instinct and so on) or using matter (where instruction is carried by arrangement on chemicals such as clay, papyrus, ink, *DNA* or other messengers). Storage may be fixed (as in a file or photograph) or dynamic (as in a running film, program or automated mechanism).

Causal reason generates meaning. *All active information is meaningful.* Implementation of a plan involves *semantics whose specific meaning transcends the generality of grammar and syntax.*

A command erases randomness; code is a deliberate restriction imposed to cause a non-random outcome. It employs an agent of restriction - sign, symbol, code of one sort or another. *This is the world of signals, semantics.* **Whatever is encoded is intentional.** Coherent code and incoherent chance are chalk and cheese.

While, to cut the story short, *the upper level of code is its abstract infrastructure, the lower is its physical expression (in chemical or energetic e.g. vocal form).*

↓ *irrationality*	*logic* ↑
no-message	*message*
no-code	*code*
physico-chemical maelstrom	*psychological scheme*
accident	*teleology/ purpose*
chance construction	*technology*
mindlessness	*mindfulness*

Top-down, therefore, chance loses out. Code is always the result of a mental process. If a basic code is found in any system, you might conclude that the system originated from a mental concept, not from chance. You might therefore conclude it had an intelligent source - especially if that code is optimised according to such criteria as ease and accuracy of transmission, maximum storage density and efficiency of carriage (such as electrical, chemical, magnetic, olfactory, on paper, on tape, broadcast etc.) to its recipient; and if, above all, it works and orderly instructions are unerringly responded to.

A coder takes no chance. Randomness is eliminated. By definition, mistake or randomness degrades information; and the job of any editor is to eliminate interference, 'noise' or mistake. Chance neither creates nor transmits information. On the contrary, accidents always (unless accidentally reversing a previous degradation) degrade meaning and, by degree, render information unintelligible.

A coder takes no chance. Randomness is eliminated. A compiler is a mindless mechanism but a programmer is not. He determines the code and its operation: error is rigorously debugged. By definition, mistake or randomness degrades information; and the job of any editor is to eliminate interference, 'noise' or mistake. As we see, chance neither creates nor transmits information. On the contrary, accidents always (unless accidentally reversing a previous degradation) degrade meaning and, by degree, render information unintelligible.

Moreover, if information *is* immaterial then a frame (such as the holistic one that I propose) is needed; the metaphysical needs be included with subsequent shock of theories of Source, Origin or Potential to materialistic physics, psychology and biology.

All active information (whose psychology we'll address today) is meaningful. Implementation of a plan involves *semantics whose specific meaning transcends the generality of grammar and syntax.* We need frameworks (logical systems/ hardware) within which to manage information (software).

Code is devised and stored in mind but information's physical expression obviously employs material arrangements. **The alphabets of such code include, as well as humanly devised systems of communication, bio-chemicals (supremely, *DNA*), binary nervous code and the sub-atomic elements, forces and atoms of physics and chemistry.** The former couple are specific to life forms - possibly, at least with respect to *DNA*, anywhere life may exist in cosmos. The latter also constitute a universal code. This code dictates, through the agents that a study of phenomena elucidates, the way things naturally turn out. (*Looked at this way cosmos is a Grand, Dynamic and Encoded Text*).

The lowest level has, therefore, already been explained. This is the part that bottoms out in light, sound, fluid patterns and in crystalline solidity. This is the 'external', 'objective' or physical side of the psychosomatic border; and, while forces and atomic particles comprise its primary expression, the hard, bulk universe that we survey is its secondary. *At this level we find data items, that is, materials on which arrangement is imposed.* The phase appears as an object, event or reflexive pattern of behaviour without subjective quality, context or intrinsic meaning. It is the level of raw data *per se*, that is, objects in whatever form they appear *before* higher inspection, manipulation or interpretation. No subjectivity and only husk-like objectivity remains.

Now: briefly, a recap of that vital psychological invention, codes, and coding's physical expression.

Finally, therefore, let's examine three lowest level but still complex expressions of purpose. Verbal, musical and mathematical codes have produced the glories of humanity.

Music.[15]

The old word for integrated order is harmony. **Music, like health, is harmony in action**. It is an archetypal formulation of energy, constrained only by its type of instrument, harmonics and the skill of its musician. Melody is a most profound form of information-in-motion. It is perhaps the best medium for the vibratory transmission of meaning. Thus, regarding composition, First Cause does not need brain!

discord	*harmony*
disease	*ease*
blockage	*flow*
incoherence	*coherence*

Logical Archetype, primordial vibration's song, will do! Symphony composed by resonant association is enough! The vibratory energy of First Cause and its subsequent constructions would embody the internal logic, patterns or rhythms that pulse through each level of the grid of creation. **Therefore, beyond the other cosmic models and although it can't be posted in a diagram, music is Natural Dialectic's Master Analogy.**

Machine.[16]

The table you are sitting at is a device of simple and unmoving parts. Its construction serves a purpose and, therefore, has 'mind in it'. Its shape is 'passively informed' from chemicals that would never take that shape without direction. In fact, every organism marks the world with trivial, crude or intricate constructions for survival and, for some, passion's further purposes. Look around. This room is everywhere imbued with mind. Physical and chemical analysis, howso exhaustive, will never of itself find mind and, therefore, reason in its material parts. Of course, an investigator recognises this pair is present but cannot calculate it or discover any mental particle. He cannot, therefore, in a strictly scientific sense, complete his explanation for a table or for any other object, such as a computer, in this space; in fact, his report must forever miss the main point, that is, their purpose. Spread your wings. Outside the room a town is bustling. Cambridge, with its streets and buildings, is a human mind-world. **The hidden factor in plain sight is information.** Everywhere you easily detect informants and informed. Why, therefore, is information 'outside science' just because it's metaphysical not physical?

Not physical? To lasso the intangible and corral psychology within the ring of physicality has been assiduously addressed. *This address is an attempt, using several strands of argument, to materialise the fact of information.* In the next Lecture we shall examine the idea that chemicals can produce conscious experience, that is, scientifically brain *is* mind and thus our personal information-centres are material alone.

"Machines", argued philosopher/ scientist Michael Polyani, "are irreducible to physics and chemistry." They are irreducible because they involve immaterial purpose, the stepwise development of a plan of implementation, a directed cohesion of working parts and, of course, the thoroughly non-material anticipation of an operational outcome. Such a machine may itself be simple or complex; it may be possible to deliberately adapt its purpose and therefore function; but if parts are missing or corrupted then its operation fails. *Engineers routinely face the chicken-and-egg problem of designing interdependent components for their mechanisms and machines.* They are not inclined to leave their solutions to chance. Thus, although a machine is totally material, it is at the same time full of its inventor's mind. It has no mind but, paradoxically, by proxy does.

Let us rephrase. *A machine involves a system of well-matching, interacting parts that, unless any is removed or degraded, contribute to a function or produce a targeted result. Such systems are therefore specifically and irreducibly complex; and so to work they must be made at once.* In this case which comes first - a machine or its concept, a work of art or its inspiration, the chicken or its egg? Extrinsically fruit issues from the branch. The branch came first and the fruit after. *From an intrinsic point of view, however, the branch is from the fruit. The fruit came first but, in order to bear it, the branch was conceived.* **Wherever an apparent 'chicken-and-egg' situation crops up the puzzle is resolved by the introduction of purpose**.

Let us also be absolutely clear that, although machines are operationally subject to the constraints of natural forces and environmental context, they are never created by them. *No machine ever appeared as a consequence of the addition of random free energy* *into an existing system.* **Physical nature can't create machines; yet information and machinery are closely twined in living and, indeed, all teleological constructions.** *You want evidence for holism. This is fact.*

Mind Machine.

A computer is a mind machine. Inspect the logic of its functionality. Examine integrated circuits. Their molecules, like those of a brain, show no sign at all of mind. They are not even biochemical but metal, plastic, silicon and so on. Yet they are replete with passive, rigidified order. In this respect each one's determination cries out its ghost, the active order of its maker. The whole machine is absolutely full of maker's information.

Programs, coded programs are a mind machine's intent. They express the will that mind invested in machine. *It needs be re-emphasized, every machine (including body biological) has the mind of its maker in it.* Not in its atoms *per se* but in its purpose, design and lawful operation. Machines passively embody information. Mind-machine information is passive. Active has produced passive information. **There is, to re-emphasize the fact, known neither law nor process of nature by which matter originates information.**

What, therefore, might we deduce about the presence or the absence of a Maker of the Universe?

Lecture 3: Psychology

Bear in mind, as we approach psychology, that Natural Dialectical Philosophy *is* internally self-consistent. Let's equally and importantly remember that, as in the finger-waggle, its knock-on networks of causation are not simply 'horizontal' (as with one-tiered materialism) but its 3-dimensional structure includes 'vertical' causation.

Lao Tse is seen here with Swiss psychologist Carl Jung. **Tse's cosmic fundamentals and the basic form of Natural Dialectic (the dynamic interplay of complementary opposites) are illustrated here.** Jung was deeply influenced by the Taoist I Ching or Book of Changes.

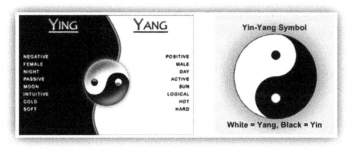

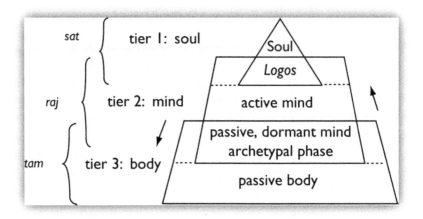

tier 3/ tier 2	Tier 1
body/ mind	Psyche/ Soul
lesser selves	Transcendent Self
conscio-material range	Conciousness
physic	metaphysic ↑
energetic	informative
material	immaterial
non-conscious	states/ grades of mind
body/ tier 3	mind/ tier 2

With the idea of a 3-tiered universe do you also recall the logic of a monk or yogi, a contemplative – man microcosm of the macrocosm?

In meditation they seek to remove the existential motions that arise in body (matter) and in mind. Once the waves of mind are tranquillised they no longer obscure the Central or Peak Truth.

This is *materialistic nonsense* since thought and subjective experience (by sense or otherwise) are things. But could senseless atoms ever yield their own experience? Or mathematics? Reasoning? Morality? Intelligence?

Remember that the world, for a materialist, is basically colliding particles and interactions due to force. And, biologically, neither cells nor their nucleic acids know or care a fig. Such oblivion is ignorant of laws of logic, creativity or purpose. Why should compound intricacies of atoms, up against the laws of entropy, codify for algorithms and the mechanisms for an intricate survival? Even *if* blind carelessness developed nerves (a great guess in doctrinal dark) why should their chemical reactions produce a calculation or a sonnet? *How can a set of physical changes, physically caused, possibly 'correspond' to such conscious experience as seeing that 'an axiom is self-evident' or to conscious logical transition as implied by the word 'therefore'? If a whole sequence of logical steps, mathematical algorithms or reasonable statements were indeed merely the effect of a causal chain of physical processes, all blindly and mechanically or electrochemically determined, it would follow that the speaker could not think what he wanted or help saying what he did. He would, quite literally, not know what he was talking about. His statements, as reasoned arguments, should therefore carry no weight. Thus, why should we believe a word he says?* **In this respect it is common knowledge that, on his own terms, a serious materialist is talking gobbledegook.** Physicist Sir John Polkinghorne rams the point right home. 'Thought is replaced by electro-chemical neural events. Two such events cannot confront each other in rational discourse...The very assertions of the reductionist himself are nothing but blips in the neural network of his brain. The world of rational discourse dissolves into the absurd chatter of firing synapses.'**

What about a sense deprivation tank? Does the experimenter's mind disappear? What about the prisoner in the isolation cells at Alcatraz? Or you? Is thought an object you can analyse, like nerves, underneath a microscope? Can you measure your own thoughts; can you precisely picture, measure or objectify a feeling's subjectivity? If not, then perhaps, like numbers, they are immaterial (though most important to you personally). If the key element of mind, your subjectivity, has simply been reduced to ghostliness and the central, immaterial dimension of psychology's reality electrified then how far past a neurological glass-ceiling has science scaled?

Please, therefore, allow at least, the axiom that information, active or passive, might exist in immaterial form; that mind is metaphysical.

The word *'psychology'* means study of *psyche* (from a Greek word generally translated 'soul'). But current 'psychology' were better termed *'noology'*. This is because the discipline studies mind, for which the Greek is *noos* (from the verb 'to notice' or 'to think'), and not *psyche*. Or does it? Perhaps not even 'noology' goes far enough. **Scientific materialism can, by definition, only allow physical composition.**

So what does the word 'soul' or 'psyche' mean?[17] Inner essence? Centre? Does it include immaterial consciousness? *Bottom-up*, it is figment of brain's motion, an imagination born of nerves.

Top-down, **however, might you not reflect creation as a whole? Is your own constitution, microcosm of the macrocosm, in the image of a three-tiered universe?**

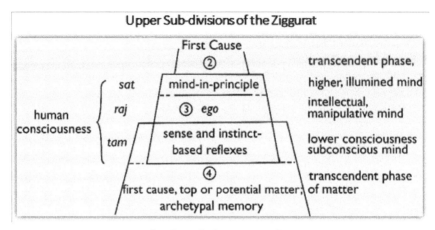

Upper Sub-divisions of the Ziggurat

Hierarchical Psychology; Mental Ziggurat

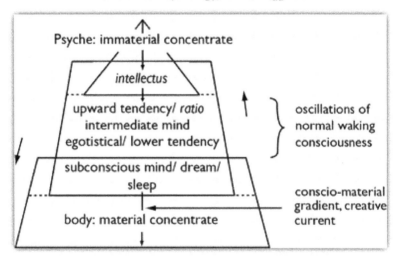

In this analogy the anti-parallels run *in* (for understanding), *out* (for action's creativity); input, output - their two tracks lead to and from an information-centre at the head. In this respect they form an image representing you. This is how, in body's tier-3, passive zone, your nervous system is arranged; *indeed, as we'll see, it represents the informative way that your whole body is disposed, that is, it represents the structure of biology.*

At this point it may also be noted that, from a *top-down* point of view, created 'mind-with-body' is seen as a 'soul vessel'. Thus every different pot (or organism) contains the same pure water. In this manner a human may be seen as soul having a physical experience rather than a body erroneously imagining its metaphysical soul.

Clearly, we have two very different views of mind and soul.

Substitution, metaphysical for physical, is, however, scientifically 'unacceptable'. 'Impossible' - whatever future studies might disclose.

'Immaterial' is a word too far, a taboo; it breaks the mind-set's basic rules; it kicks a prop that's critical to sustenance of exclusive, naturalistic faith. *So, creed decrees, thought's entirely a nervous matter.* **Thus, naturally, psychology emerges as the study of neurological phenomena.** Carbon, oxygen and more, from this perspective soul is the activity (incredibly, incomprehensibly complex, mind you) of particles. Consciousness 'emerges' from non-conscious molecules grouped in some special way. Isn't thinking generated by the soft-wired workings of a brain? *Just sling sufficient atoms, in the form of nerves, together - they'll become no less than self-aware!*

That mind is a physical illusion is the neurological delusion.[18] **To mistake its neural correlate (as, say, registered in brain scans) for experience itself is an error as basic as taking the electronic pulses in a wire for all there is to telephonic conversation. It is a prime, elementary fault, a first category philosophical mistake; it might be termed full-blown, psychological mythology.** To identify consciousness as an illusion is itself, denying the reality of one's own experience, a pernicious - even dangerous - delusion.

One party, it is clear, *believes life is a phantom of the atoms.* Brain *causes* subjectivity. Thought (therefore belief and all the purposive effects of will and faith) is part and parcel of nerve chemistry. And what is the *experience* of consciousness? The essentially robotic view of neuroscience holds that nerves *are* consciousness. We just don't yet understand, the faithful purr, how brain's 'emergent properties' can squeeze experience out or how the juice that's 'you' must be exuded from its molecules. A revelation is, however, prophesied. Materialism's scientific certainty decrees that life will be reduced to chemistry and mind experimentally identified as simply due to complicated ionicity! You are a product of your physiology and so, at root, your genes alone. Life has, hasn't it, to be an electronic after-thought?

Nervous particles and atoms aren't, like atoms anywhere, alive. Therefore, if life is made of them it shouldn't be alive! *Thus the other line suggests, conversely, that brain isn't an originator but a mediator.* It is our *dashboard* as we fly through life. A filter. A sophisticated interlink between mind, body and the latter's physical universe.

If so, it is a **transducer device** that, like any mediation network (e.g. radio), must be sufficiently well-constructed to handle large volumes of two-way traffic. It accepts environmental signals and translates them ($\uparrow$) 'upward' into mind's experience; and issues orders ($\downarrow$) 'downward' into body chemistry. As an organ of 'cockpit control' its 'dashboard' accurately connects an immaterial mind to a material body and, thereby, physical conditions. Of course, young pilots (babies) have to learn to fly; thence we and other kinds of creature navigate, in the vehicle of body, various sagas on the senseless stage of matter. In this view mind and brain, although compounded, are quite separate entities - the former metaphysical and latter physical. Brain chemistry's identified as a design that expedites exchange of information. Your head is thus a medium!

Making no material difference by adding immateriality, the Dialectic simply reconstructs creation on the basis of a 'conscio-material' duality. In short, perhaps brain neither does nor ever did enjoy a seamless, subjective experience. *The implications of this seminal idea are so extensive that this whole course is exploring them.*

So now to <u>Consciousness</u>.[19] This is what it's all about. Without it you are nothing. The star of every play is mind; the kingpin of psychology is consciousness. What is the 'thing'?

For Natural Dialectic there's an immaterial element of information. This element of knowledge isn't physical. Non-conscious matter is a special case of its subjective absence. *It is pure non-consciousness.* Gases, streams and solid bodies don't know anything. Their oblivion's polar opposite is total wakefulness. Of what, you ask, does this consist? As matter's pure non-consciousness exists could not a concentrate of immaterial information - pure consciousness - have being too? So that creation's root turns out to be oblivion's antipode - *Potential Knowledge, Latent Field of Knowing, Pure Consciousness.*

Materialism doesn't like this sort of phraseology at all.

Essential Psychology

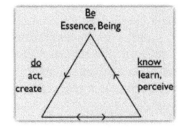 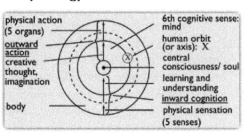

do/ know	*Be*
objective/ subjective effects	*Subjective First Cause*
lesser levels of consciousness	*Pure Consciousness*
body/ mind	*Psyche/ Soul*
passive/ active expressions	*Potential*
↓ *passive information*	*active information* ↑
programmed	*programming*
non-conscious/ objective	*subjective/ conscious*
reflex	*voluntary*
physical action	*psychological oversight*

<u>Hierarchical Psychology</u>. *Top-down*, any attempt to reconcile science and psychology must logically start at the top (Conscious Soul) rather than bottom (matter). It must start at Axis, Centre, Source or Pivot

48

rather than with the effects of subsequent informative or energetic motions. Therefore we deal first with the two conscious states, super-state transcendent and our ordinary, restricted awareness. Then we fall to cover, at the sub-conscious end of mental balance, the conditions of dormant mind, dreaming and deep-sleep; finally we inspect the psychosomatic domain of instinct, personal memory and archetype. The latter account for the psychosomatic connection between mind and its sub-state, non-conscious material body.

Top-down, therefore, if non-conscious matter forms the cosmic sink its source is consciousness. *Uncreated Consciousness is primary and existential forms are secondary*. **What light is to physics Essential Transcendence is to psychology; in other words, the only absolute measure of consciousness, and therefore psychology, is Transcendent Super-Consciousness**. From mind holistic distillation therefore reaches high; but materialism recognises only base degree, the lifeless one.

range of consciousness	*Super-State*
normal states	*Ultra-/ Super-Conscious*
degrees of obscurity	*Clear Mind/ Transparency*
partial waking	*Fully Awake*
↓ *inattention*	*focus* ↑
blinder mind	*clearer mind*
unconscious/ oblivious	*aware*
infra-/ sub-conscious	*conscious*
involuntary	*voluntary*
sub-state/ unawake	*partial waking*

Five Main States of Human Mind in Relation to Brain

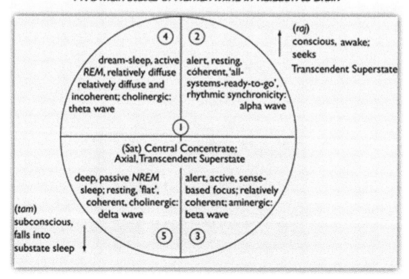

The stack illustrates triplex psychology. Its three states are Super-consciousness, normal waking and sub-consciousness. And, as immediately above, we can expand these to relate five main states of mind with their correlated conditions of brain.

Of these states, as also shown in the stacks below, 1 relates to Super-consciousness, 2 and 3 to waking and 4 and 5 to sub-consciousness.[20]

Vectors and States by Stack

info. out/ info. in	*Information Centre*
lower/ higher	*Third Eye*
↓ *lower*	*higher* ↑
sensory bias	*contemplative bias*
spiritual anaesthesia	*material anaesthesia*
diversification/ detail	*unification/ principle*

spectrum of consciousness	*Super-Conscious*
relativity of forms	*Centred Meditation*
(2),(3),(4),(5)	*(1)*
↓ *downward tendency*	*upward tendency* ↑
passive/ exhausted	*active/ alert*
involuntary	*voluntary*
non-conscious/ subconscious	*conscious*
busy-in-sleep/ dreaming/ theta (4)	*busy awake beta (3)*
deep sleep/ coma/ delta (5)	*contemplative/ alpha (2)*

Libraries have been written concerning the state of psychological normalcy, our waking state (here 2 and 3).[21] Suffice here to note that *ego* is a mask, per-sona, prosopon or prosopeion. Its dynamic structure, necessary for function in a body, amounts to a structure, face or formful covering of inner, underlying consciousness. It is this reducing agent that a contemplative attempts to 'vaporise'.

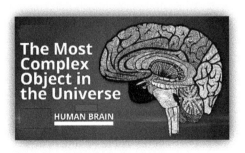

We are now in a position to connect metaphysical with physical hierarchy, the latter as it appears in the construction of the most complex object in the universe - a human brain. **Claimed to have evolved out of a total lack of logic it can, however, produce a logical and comprehensive understanding of itself. Boot-strap logic here,** *par excellence*!

Here is brain drawn simply.[22]

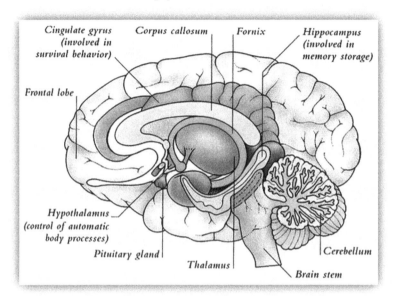

Top-down, **on the other hand, brain is seen as a mediator, a fine linkage factor between metaphysical mind and physical body to which, for a limited period, it is attached. The brain is a machine, a mind machine like a computer, and this is its task.** In such context the next three illustrations represent its dialectical construction. The third shows a nested version of the ziggurat (Brain Very Briefly) looked down at from above. Take a little time to work out what, *top-down*, is being said.

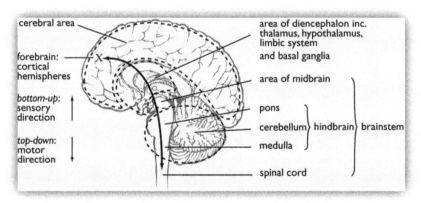

These dialectical drawings of the logical, top-down construction of the brain, are at the same time an organic representation of the levels of mind.

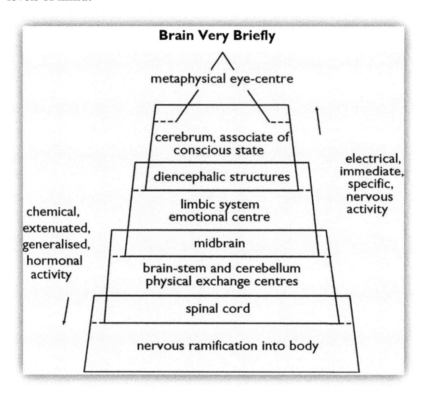

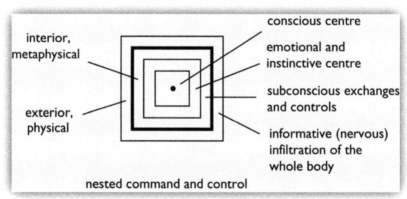

Is there no logic in the way a brain is built; or is it just an aggregate of happy accidents? *It is logically surprising that, for no reason, non-conscious and illogical chaos should have constructed order to a very highly systematic climax in the most complex working system of the cosmos, an information processor whose whole, sole,*

negentropic business is order - a central nervous system and associated brain. Did matter, getting far more than it didn't bargain for, perchance 'evolve' a brain? Materialistically speaking, it must have. *In reality did it?*

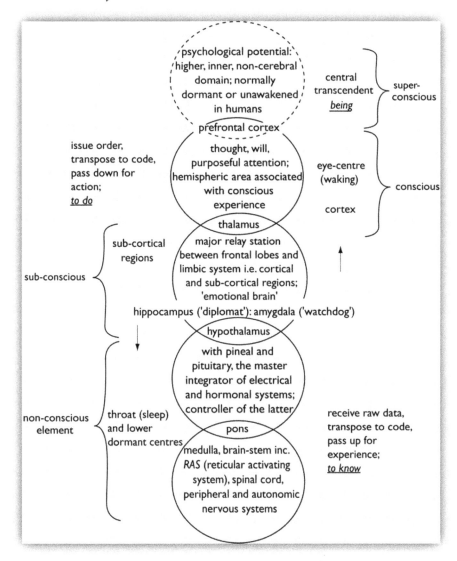

A brain so simple we could understand would be so simple that we couldn't; but, *top-down*, we can enumerate the principles round which its great complexity might gather.[23]

The next diagram confirms a polarity-based construction of the brain.[24] **It shows the hemispheres dialectically (along with Roger Sperry, doyen of split brain research). As such** illustrates how the lateralized or polar construction of cerebral cortex accords with Natural Dialectic; it includes the mind-to-matter, informative-to-energetic or brain-to-body inversion.[25]

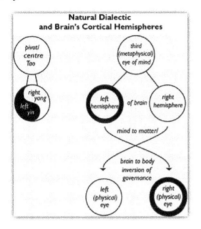

Finally we note Quality of Information and, therefore, the natural ascent of knowledge-seeking man towards Transcendence, towards his Essential Centre.

Quality of Information

lesser truth	*Truth*
shades	*Light/ Consciousness*
grades of knowledge	*Knowledge*
grades of priority	*Importance*
↓ *negative*	*positive* ↑
error	*truth*
dim	*bright*
oblivious/ ignorant	*knowledgeable*
increasingly misinformed	*increasingly informed*
incorrect/ wrong	*right/ correct*
trivial	*important*
inefficient	*efficient*
useless	*useful*

Information is often called by its users 'intelligence'. *In fact, intelligence is a function* Life is intelligent, always goal-oriented (teleological) and orders things according to its designs. *The greater an intelligence the faster, more accurately, intricately, thoroughly and efficiently are its problems solved and complex goals achieved. In short, the more teleological is its nature.*

The importance or triviality of information is rated relative to various desires and sets of priorities. Such priorities and the wishes that

54

sequenced them change. Instinct, education, advice and cultural experience help shape such changes and, in life's relative muddle, provide answers and directions.

Thus, although everybody operates according to their own variable, personal set of priorities, do there also exist permanent, natural and universal priorities? Is there a hierarchy of information, a set of principles? If so, what is the quality of nature's 'top set'? Is there Prime Information? Is there anything of Absolute Importance?

The Ascent of Man

existence	*Essence*
lesser being	*Supreme Being*
expression	*Potential*
body/ mind	*Soul*
periphery	*Centre*
↓ *descent*	*ascent* ↑
compress/ diminish	*grow/ liberate*
passive/ involuntary	*active/ voluntary*
created form	*creativity*
materialisation	*dematerialisation*
entropy	*negentropy*
devolution	*evolution*
body	*mind*

From high to low, we now descend to the subject of *sub-consciousness.*[26]

The Sub-conscious, Psychosomatic Sandwich

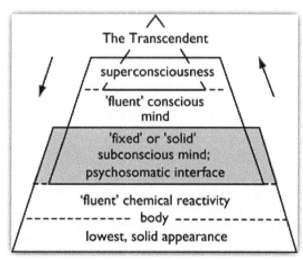

55

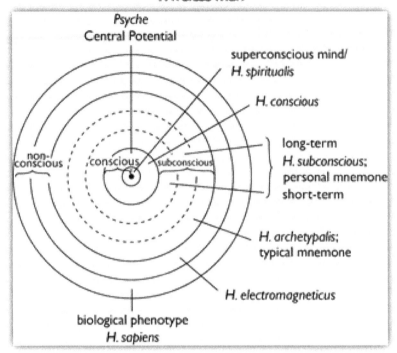

Are you ready for a fall to the diffuse conclusion of psychological entropy, for a subjective drop into the labyrinth of underworld? You know what it is to be mentally as well as physically exhausted. You've often dropped off into mind's flat, dark condition we might call inertial equilibrium. 'Little death' is not the world's end so let us take a snooze cruise; it is time to fall asleep.

Dreaming. For dreamers dreams are real enough; but the experience is untrammeled by either external events or the ability to reason. Waves wash equally on what is in their path; a torch shone randomly around picks out disconnected or illogically connected objects and events. The files are scattered, narrative is blurred.

In this apparent chaos is there reason? Dreams serve, like waking mind but lacking sensory restraint, to relieve circumstantial pressures in the form of dangers, problems and anxieties. Subjective equilibration is the game. For relief's solution, relaxation and release of tension, dreams roam uncontrolled through memories. Of what is lunar dreamscape made but these? We'll soon explore the subconscious world of *personal and typical (instinctive) mnemones,*[27] that is, major files of memories. Such world is not *per se* irrational but, carried passive on an incoherent dream-stream, an observer's various slumberous visions of it is. Sometimes, waking, he recalls his lucid travel or more jumbled narrative. There are even, in the loop, memories of dreams.

Deep Sleep. In deep, non-*REM* (or *NREM*) sleep the 'upper', voluntary structures of brain are cut from the loop. A sleeper's movements, including eye movement, are much restricted; sensation is dull or absent. Brainwaves, the overall coordinators of the central nervous system, slow to between 0.6 and 3 hertz. These are so-called delta waves. Maybe deltas drop to zero. Brain death. If, by head injury, stroke, tumour or poison, the sleep/ wake toggles fail or signals cannot reach the forebrain then the patient drops into oblivion. The curtain falls but drama does not start again. Coma is an open tomb, an unpinned shroud or wake-less sleep - though in its stillness deeper grooves of mind (archetypal constructions but also profound personal impressions and rote such as language) stay frozen yet intact.

Organisms sleep in different ways and yet, not dropping off into their 'little deaths' at all, they'd die. Sleep is vital. Survival is insistent, for example, that a brain is regularly cleaned (for example, sleep's neural shrinkage lets in spinal fluid to wash toxins out). How, though, did genes evolve the physiology for 'off-line' maintenance or *REM* and *NREM* sleep? How did mind evolve from matter? Because, as well as chemical complexity, dormancy's a metaphysical affair; it knits up *mind's* (not brain's) ravelled sleeve of care. A just-so story for such wonder, best beloved, is that elevation from the 'psychology of physic' into coma must have woken life from its primordial lifelessness; and that, before you sleep on it, evolution really woke up when, atomically, the abyssal chemistry of slumber re-arranged itself as conscious beings such as you!

Frozen Time. You sleep but your past does not disappear. You wake and your past has not disappeared. You think you have forgotten, you may even suffer amnesia but untapped memories remain. They are how we freeze time. **A memory is frozen time. It symbolically encodes the past.** *A memory is a thought object and, as such, has no life of its own.* A disc encodes music once recorded 'live'; it's a memory that, when replayed, affects the present and, from this, the future. Thus mental memory is an encoded image; it is a record and, on conscious recollection, becomes a presence of past action that may affect the future.

***Top-down*, neither memory nor knowledge are inherently physical.** No doubt, correlated nervous circuitry acts as a storage-and-retrieval system that, by association, allows the immaterial library of remembrance its efficient, selective interaction with an innervated body and its circumstances. Thus 'engrams' (are they sited at synapses, inside a nerve cell's body or elsewhere?) may, if they exist, indeed *relate* to physical experience; they may act as a recognition trigger, reference point or body's resonance with an experience. And organs (such as hippocampus and amygdala) certainly seem to log experience in the manner of a record/ playback head; they catch or release a moment that, in fixity, is called a memory. But if such 'storage' or 'playback' button fails the system's compromised. Either records are not made in the first place or the connection becomes impaired or irretrievable. *But the 'disc' of memory itself is metaphysical.*

Memory, the only form of metaphysical information storage, is the shape of infra- or sub-consciousness. *Indeed, it is sub-consciousness.* Subconscious mind consists of coded files that we call memories. **It is made of memories**. Regarding life-forms memory comprises an organism's library of precondition and conditions, that is, its context for experience. The precondition is its archetype, the basis of its sort. We might call this sort of memory *'typical'* while *'personal'* experience includes both active (created and transmitted) and passive (received) information. Such memories are not necessarily 'frozen' like a photograph.

Nor, although it comprises a concept or basic expression of an idea, is an archetypal memory. Memories may operate like movies and store programs that, once triggered, can unfold like stories in a sequence to their completion. Any stored plan is, like a film or computer program, such a memory. Programs are, although dynamic, still a frozen form of mind; and they're replete with information. They specify the most efficient means to a well-defined end. **You might argue in this vein that biological structures are codification incarnate; and that the concept they express is an archetypal program**.

Archetype[28] **we see as a First Cause**. There are First Causes psychological and non-conscious physical. **Archetype and matter-in-principle.** These will lead us to a triplex view of psychology, physics and biology.

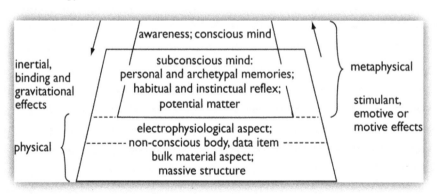

Archetype is, like any idea, the informative potential for its own later expression. The word means basic plan, informative element or conceptual template. Like instinct it is, according to holistic logic, held in subconscious mind, the subconscious element of universal mind. Thus it consists of pattern in principle; it is the instrument of fundamental 'note' or primordial shape, the causative information in nature or 'law of form'. It is nature's bauplan or blueprint. **Called potential matter its fixed files are seen as hard a metaphysical reality as, say, particles are physical realities**.

This diagram illustrates internal hierarchy and the place of *H. archetypalis* in biology.

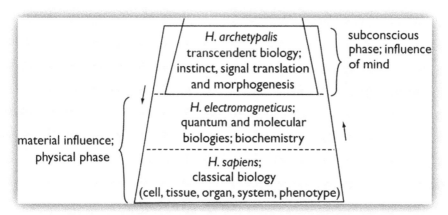

Of modern psychologists Carl Jung had perceptive sympathy with the idea of archetype. Although he did not cast his net wide enough to include all organisms and the whole inert part of cosmos he nevertheless suggested that human archetype involves elements of the 'collective unconscious' or 'universal psychic structures'. Such 'psychophysical patterns' are given specific expression by human consciousness and culture. Furthermore Jung uses the analogy of light's visible spectrum wherein normal human consciousness is yellow with unconsciousness at both red and blue ends: Natural Dialectic's conscio-material gradient likens the visible spectrum to normal mind with ultra-violet scaling super-consciousness and subconscious infra-red grading down to extra-low frequencies of non-conscious matter). Thus Jung also sees the archetype as a bridge from mind to matter. Finally, crucially, just as local objects and events are details distinct from the general principles that govern them, he differentiates between specific, local, conscious imaginations (or 'motifs') and the 'unconscious' generality of archetype from which they derive.

Not only Carl Jung but founding fathers of neurophysiology addressed the mind-matter (psychosomatic) issue in which archetype plays, dialectically, such a pivotal role.

And "What greater inherent improbability", wrote a founder of modern neurophysiology, Sir Charles Sherrington, "than that our being should rest on two fundamental elements than on one alone?"

And Sir John Eccles, Nobel prize-winner and another founding father of neurophysiology wrote, "Now before discussing brain function in detail I will at the beginning give an account of my philosophical position on the so-called 'brain-mind problem' so that you will be able to relate the experimental evidence to this philosophical position. I have written at length on this philosophy in my book *Facing Reality*. ... I fully accept the recent philosophical achievements of Sir Karl Popper with his concept of three worlds. I was a dualist, now I am a trialist! Cartesian dualism has become unfashionable with many people. They embrace monism to escape the enigma of brain-mind interaction with its perplexing problems. But Sir Karl Popper and I are interactionists, and what is more, *trialist interactionists*! The three worlds are very easily defined. ... such classification takes care of everything that is in existence and in our experience." He continues, "World 2 is our *primary reality*. Our conscious experiences are the basis of our knowledge of World 1, which is thus a world of *secondary reality*, a derivative world. ...We are all the time, in every action we do, incessantly playing backwards and forwards between World 1 and World 2."

Although he includes what he calls soul and spirit Eccles does not, however, rate them as the dialectically distinct Essential level 1. He misses, as do most of us, the Peak of Mount Universe. Thus, as regards the Three Tiered Ziggurat (Lecture 2) his World 1 corresponds to the unconscious physical base (level 3); his World 3, knowledge in the objective sense, to mind (level 2); but his World 2, states of consciousness, also corresponds to dialectical level 2.

In this case, how does mind interact, primarily at least through brain, with body? What might constitute the nature of an interface (*PSI*)? **We turn to the Dialectical bio-classification of** psychosomatic linkage.[29]

On the subconscious side of the psychosomatic interface *personal memories are viewed as programs riding on the wave of an archetypal channel* (e.g. the human channel). For a materialist, who guesses that consciousness is a peculiar effect of certain formulations of non-consciousness and, therefore, informative consciousness is an 'outcrop' of brain chemistry, the next section is obviously irrelevant.

On the other, *top-down* hand, check this algorithm, graded through synchromeshes 1 and 2, of psychosomatic linkage by domain.[30]

Psychosomatic Linkage by Domain

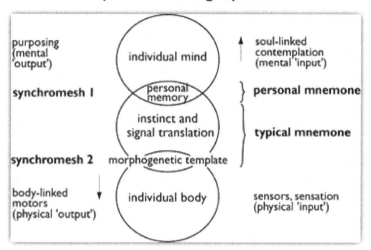

A typical mnemone is composed of three main sub-routines; or, if a protocol is a standard procedure for regulating the transmission of data between two end-points, three linked protocols.

Dialectical Bio-classification

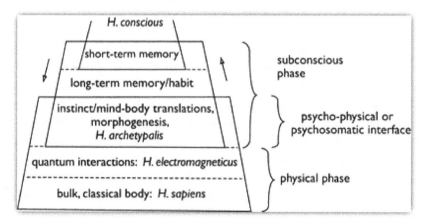

Dialectical Bio-classification corresponds with the rings of Wireless Man. In this case it classifies yourself.

Together translational, instinctive and morphogenetic programs comprise an individual's archetype. If, as in the case of plants, fungi etc., there is no conscious component, then the translator element is reduced from nervous to chemical (e.g. hormonal) messaging alone.

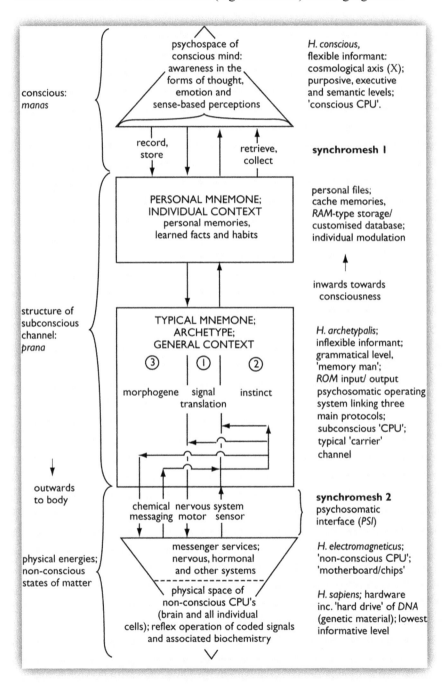

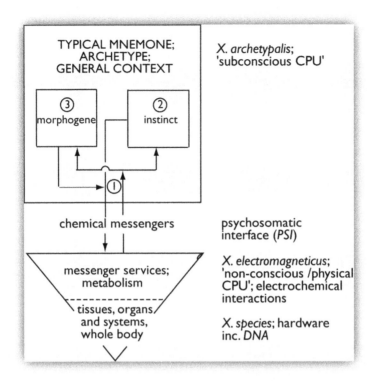

This couple of diagrams illustrates a suggested architecture of the subconscious[31] for both conscious and unconscious organism. **Every cell contains chemical genetic information; it is suggested each also acts as an antenna for its typical mnemone, that is, its archetypal broadcast.**

If you think 'archetype' is just a 'cop-out' in explaining how things orderly proceed then I suppose a systems analyst must think conceptual plans for any working mechanism or his working chips are 'cop-outs' too. However, philosophical objection to a program's purpose in no way mitigates the impact of its natural possibility and, if an immaterial element of information exists, its natural fact.

The basic principles of archetypal psychosomasis are by now clear. *Mind (at most gross, subconscious level) conjoins with matter (at subtlest, least-massive/ almost-immaterial level);* elusive quantum *probabilities pinned-down substantiate, it seems, certain processes; photon, electron and their effect on quark (protonic) or atomic position precede, in the sense of underwrite, molecular and bulk reactivities; and, where electrodynamics describes the effect of moving electric charges and their interaction with electric and magnetic fields, biological electrodynamics precedes all bio-molecular considerations.* **Every biological process is electrical; and the flow of endogenous currents is the primary and not secondary feature of physical life.**

Not only biochemistry but quantum biochemistry heave to the fore. Natural Dialectic lifts perspective from molecular to a vibratory, field perspective. *It is thus suggested that, at electrical and wireless levels, patterns of subconscious mind meet and influence matter; archetypal information is relayed to chemistry by polar charge and light.*

H. electromagneticus[32] has, in the human case, been identified as the physical side of message-exchange and *H. archetypalis*, with its routines, the metaphysical correlate.

In 1935 Dr. H. S. Burr, Professor of Neuroanatomy at Yale University, and Dr. F. S. C. Northrop established that all 'living matter' from a slime-mould to an elephant, from a seed to a human being, is surrounded and controlled by electrical fields. Indeed, every cell pumps ions to maintain a healthy, electrical potential across its lipid membranes.

Electromagnetism involves photons. Is light not in fact the language of life's cells? Fritz Popp, founder of the International Institute of Biophysics at Neuss, Germany, proposes that low-level light emissions are a common property of all cells. Such weak luminescence ranges from thermal (infra-red) to ultra-violet. Moreover, electrochemical forces across cell membranes help control their permeability. Popp's colleague, Marc Bischof, believes that weak, coherent e-m fields combine to regulate not only the cell's surface but its internal members. Thus, correlated with the positions, densities and movements of electrons, a 'signal web of light' might harmonise cooperation of organelles with each other and with chemicals throughout the cell. Moreover cellular cytoskeleton's coiled, semi-conducting filaments and tubules, whose other jobs include structural support and transport track-ways, compose a network for the conduction of charge and generation of electromagnetic fields.

These 'wires' of electrodynamic propagation may power up other structures such as protein alpha-helices and coiled/ solenoidal *DNA*. Indeed, Bischof postulates that *DNA* pulses as a 'light pump', that is, as if an aerial both emitting and absorbing light. In short, **radiant systems, whose weightless quanta represent the purest form of physical energy, are logically identified as a prime candidate for the physical side of the psychosomatic exchange of information.**

There is no doubt, the universe is in vibration. The cosmic transmission of information and energy is, at root, vibratory. **Ordered oscillation is called harmony; harmony is the grammar of music and music is a universal language**. The theory of music is implicit in any recital. Could it transpire that explicit order of a cosmic recital is the product of implicit notation? Could its excitement represent cooperative forces, specific 'notes' called particles and thereby, all in all, harmonic code?

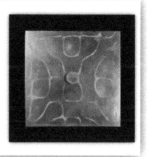

Here we note the work of, among others, Ernst Chladni (above) and Hans Jenny. **Their study, now called cymatics, involves the use of wave frequencies to precisely control the production of shapes in water, air and other media.** it is significant that Chladni's figures often imitate familiar organic patterns that we see in nature and, especially, biological structures.[33]

Pattern clearly relates to frequency of cycle. And resonance, whose orderly aspect is characterised by an analogy with music, is the tendency of a body or system to oscillate with larger amplitude when disturbed by the same frequencies as its own natural ones. *Cymatics therefore intimately involves the vibratory transfer of energy.* Such transfer is an integral part of all vibratory systems wherein waves interfere with/ destroy or cohere/ amplify each other. Energy is amplified and transferred by resonance and attunement. Common examples of electromagnetic resonance include tuning a transmitter/ receiver and photo-electric initiation of the photosynthetic process, that is, the first step in life's chemistry.

For quantum physics matter is certain vibratory frequencies of energy; and, from a dialectical point of view, it is simply stresses, strains or tensions in the medium of their absence, that is, nothingness. **There is, however, nothing random in a highly orderly creation derived from**

first acoustic principles. Oscillation between polarities, cycles, vibratory rhythms, <u>the interrelationship of waveforms and complementary resonance are at the heart of Natural Dialectic</u>.

And it is suggested that a key phrase in the suggested explanation for the wireless, psychosomatic transfer of information between subconscious structures of the mind and the physical plane is '<u>resonant association</u>'. *The modus operandi of psychosomatic broadcast is therefore, in a word, attunement. Resonance.* It involves transduction between recorded information (memory) and physical energy.

The Logic of Embodiment

classical/ quantum matter	*Potential Matter*
classical/ quantum biologies	*Conceptual/ Archetypal Biology*
outworking	*Plan/ Morphogene*
informative agents	*Informative Source*
↓ *H. sapiens*	*H. electromagneticus* ↑
classical biology	*quantum biology*
gross constraint	*subtle motions*
fixture/ container	*biochemistry/ drive*
structure/ anatomy	*physiology/ function*
external appearance	*internal informants*
informed product	*informant messaging*
seeming sameness	*rapid (inter-)actions*

physical/ objective correlates	*metaphysical governor*
non-conscious effects	*sub-conscious cause*
classical/ quantum matter	*potential matter*
material expression	*archetypal memory*
physical cause/ effect	*psychosomatic medium*
sensation/ action	*psychosomasis*
physical body	*morphogene*
H. sapiens/ H. electromagneticus	*H. archetypalis*
↓ *informed classical level*	*informant quantum level* ↑
chemical receiver	*radiant influence*
anatomical shape	*electromagnetic/ biochemical patterns*
H. sapiens	*H. electromagneticus*

The *third* component of a typical mnemone is the **morphogene**. It is suggested that, in concert with the *DNA* solenoid, other biochemical spectra are intimately linked with growth, development and morphogenesis. **This stack illustrates The Logic of Codified, Programmed Development.**

Our human extent can, according to the cosmic fundamentals, be illustrated by two final slides.

Human Extent: the Conscio-material Gradient

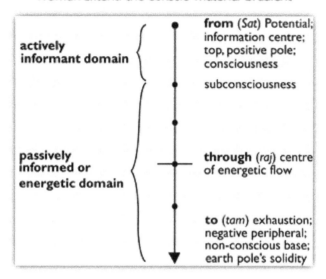

Information Man

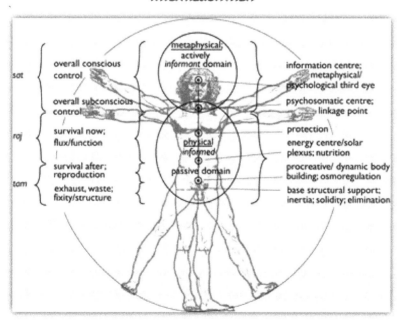

Here, as in all biology, reason's nowhere absent but, with chance, like chalk and cheese. Reason is meaningful, chance meaningless. *Why should a highly rational system irrationally 'self-organise'? Chapter 15 will demonstrate the fallacy of a belief that any chance whatsoever could invade the initial, entirely purposeful construction of a cell, a human or any other type of biological body. The chemical evolution of proto-life is a fantasy. And your own human microcosm underlines the utter feebleness of evolutionary explanation* If materialism's rational it spotlights how irrational, backing chance as its creator, rationality's become.

Isn't it, on the holistic hand, inconceivable that such a logical, integrated, self-consistent embodiment as yours, constructed with highly specific complexity in accordance with grades of the conscio-material gradient of creation itself, occurred by chance? **If reason wins, whose archetypal program is worked out in every bio-form, then chance and time's grand theory crumble back to their home ground - they bite their progenitor, the dust.**

To summarise the section on sub-consciousness, mind is linked to matter by a wireless anatomy whose instrument is the quality of vibration called resonant association.

Lecture 4: Physics

Friend Lao Tse now with Niels Bohr (the quantum theorist who, having showed that the electron orbits of an atom explained its chemical properties, neatly linked physics and chemistry) remind us that Dialectical foundations have stood the test of time. He incorporated the dualistic, dialectical *yin-yang* design into his family crest!

Other modern seekers after physical truth:

For example, these are Albert Einstein, Max Planck, Werner Heisenberg, Richard Feynman, Wolfgang Pauli and Max Born. Note three points:

1.　Man is an information-craver. He wants to know more and more about his phenomenal surroundings and the working of his own mind. He wants, moreover, to find patterns by which to order his knowledge to reflect as closely as possible the way in which his world works. Check boxes 1 to 3 of Lecture 1.

> **Physics, however, is restricted to answering questions purely with respect to and in terms of the non-conscious physical world in which we find ourselves.**

2.　Man wants to know it all and often, like a child, thinks he does. However, physicists are well aware how little we actually know. Although in 1900 it was thought that physics was, essentially, complete 2000 radically disagrees. Apparently we have only studied 5% of the universe, luminous matter. Current areas of attention with unsolved problems include:

> General physics/quantum physics
> Cosmology and general relativity
> Quantum gravity
> High-energy physics/particle physics
> Astronomy and astrophysics
> Nuclear physics
> Atomic, molecular and optical physics
> Condensed matter physics
> Plasma physics
> Biophysics

3.　Mankind may eventually completely understand the cosmos he lives in but, given the above, don't expect answers to every question in an hour or two! This talk fits physics, as far as we understand it, to the frame of ND. It gives fresh context and perspective. This especially applies because material science cannot reach actively informative, metaphysical but still natural elements; and it cannot include this immaterial 'missing factor' in its measurements. This applies to matters psychological but matters not one jot for accurate physical appreciation of cosmos; but since abductive logic and interpretation are required to deal with historical events, especially origins, regarding cosmos and biological (mind-matter) forms it may fall short. **Perhaps, indeed, human mind is not built to understand the beginnings but, being ever-curious, we'll work towards our best shot, that is, our holy grails.**

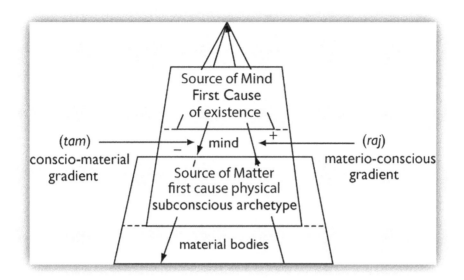

What, dialectically, are holy grails about? They are about achieving, at the heart of myriad, detailed complexity, simplicity and union. They are of two kinds - material and immaterial.

Scientists of physical phenomena and scientists of the soul travel in *opposite* directions. Their perspective differs and their goals are antipodal. But this does not mean both are not, in their way, right or that a man cannot be both scientist and a mystic.

The latter repeatedly exercise their contemplative faculty and its inward focus; they travel *up* a materio-conscious gradient away from physic towards metaphysic. But do contemplatives deserve the term scientist at all? They certainly experiment repeating a procedure to yield the single same result (and if a materialist refuses to whole-heartedly experiment, how scientific is such attitude?). This meditative grail is called enlightenment, that is, unification in First Cause. This Cause has, as we saw in Lecture 2, many names. The *Logos* is the *Tao* is The Informant. It the Most Natural Heart of Nature whereas physical archetype is the natural heart of non-conscious physic. In as much as Natural Dialectic provides a vehicle within which the causes and grails of physics, psychology and biology may be coherently, co-understood it may be called a Grand Unified Theory. Including both mind and matter it is, in effect, the *GUT* of *GUT*s.

However, scientists concentrate, as we mostly do, *down* the conscio-material gradient; they follow out through the senses and their technological extensions to learn the details of the material universe - including our own physical bodies. They seek, in a quest to find the Holy Grail of Physics, unification. This, unification (by *GUT* or *TOE* - theory of everything), is the goal with its subsequent hierarchical

organisation; but the maths of microscopic matter-in-principle (quantum theory) and large-scale matter-in-practice (relativity) do not fit together. A joiner might be **quantum gravity** and theoretical attempts to create a union are taking place. Nor are dark matter and lambda force (anti-gravity) much in the picture. It seems a Higgs boson may, at very high energies, exist but it is more hoped than clear how this tiny creature produced that most basic of physical quantities - and, if not, how did it appear? Nor, of course, has any immaterial factor ever been considered. Universal mind is not considered. **But now, in the attempt to integrate mind with matter, I want to investigate this stack of Holy Grail.**

Holy Grail of Physics	
subsequent hierarchy	*Holy Grail*
issue/ action	*Source/ Priority*
physic	*Metaphysic*
classical/ quantum	*Archetypal*
↓ *tam*	*raj* ↑
outward finality	*inner support*
aggregate/ precipitate	*influential action*
sensible matter-in-practice	*matter-in-principle*
large-scale body	*microscopic components*

Its three parts (Metaphysic, matter-in-principle and matter-in-practice) are those of the lower three sub-divisions of the following ziggurat.

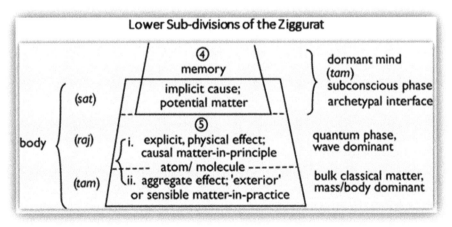

Below is an alternative version of lower subdivisions of the ziggurat. This model is interesting not only because it introduces the notion of a

<u>Triplex Physics</u>, that is, three stages of physics according to the cosmic fundamentals (*sat*) informative potential, (*raj*) action-in-principle and (*tam*) end-product, action in finished practice. In other words, it **moves towards a Theory of Physic and Metaphysic.**

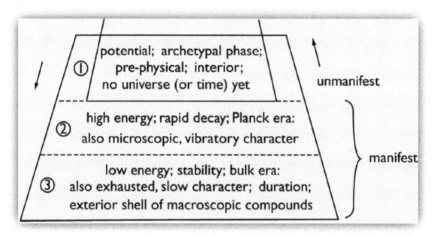

Everyone understands a horizontal, age-related chain of cause and knock-on effect. We'll discuss first causative hit, perhaps big bang, a little later; but for now, reading *top-down*, this ziggurat also illustrates physical creation in ***the vertical informative sense.*** **It starts, as you might dialectically expect, with potential or archetypal phase.** Archetype (see also Glossary) **means basic plan, informative element or conceptual template**. An archetype is the informative potential for a creation; as such it amounts to precondition. It is *first cause* that precedes action. Such action, creation, devolves from two sources, active psychological and passive physical. Archetype involves the principles generating patterns/ behaviour of mind and, lower down the scale, non-conscious matter. Archetype is a *form* or *file* of memory in dormant, universal mind. It constitutes the cosmic data bank.

The next two levels we call matter-in-principle, the staple of quantum physics; and matter-in-practice, the bulk aggregates we call the sensible universe. Let's deal with each in order.

(Sat) Potential Energy

Potential precedes possible action. It is a prerequisite or precondition for results; *but Natural Dialectic does not think of physical potential in the way that physics does (e.g. potential gravitational energy).* **Potential matter** is a metaphysical, informant whose synonym is archetype, that is, memory, code or score in universal mind.

Archetype is, like any idea, the informative potential for its own later expression. The word means basic plan, informative element or conceptual template. Like instinct it is, according to holistic logic, held in subconscious mind, the subconscious element of universal mind. Thus it consists of pattern in principle; it is the instrument of fundamental 'note' or primordial shape, the causative information in nature or 'law of form'. It is nature's bauplan or blueprint. **Called potential matter its files are seen as hard a metaphysical reality as, say, particles are physical realities**.

In the sense of *precursor* to expression archetype may be seen as a **First Cause**. Check the lower subdivisions of the cosmic ziggurat and, from implicit cause through the explicit effects of matter-in-principle to the matter-in-practice of massive bodies, the notion of a triplex physics.

An archetype is the informative potential for a creation; as such it amounts to precondition. First cause that precedes action.

Archetype involves the principles generating patterns/ behaviour of mind and, lower down the scale, non-conscious matter.

We saw in Lecture 2 (and again above) that all action, subsequent psychological and physical creations, devolves from two sources.

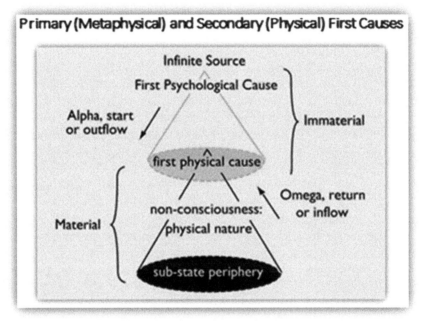

The two first causes are both informative; but the higher psychological is active and creative and the lower passive and fixed, as moulds, stencils or files in archetypal memory. This latter, first cause physical, is the 'father' of physics, the generator of precise quantum forms. In short, as thought is father to the deed or plan is prior to ordered action, so archetypes precede physical phenomena.

With or without a cosmic program, we can reasonably claim invariance under transformation. Underneath the world's commotion basic character of basic parts remains the same.

Contexts change but automated rules of play do not. *And without conserved invariance you can't obtain the balance that equations need.* **Energy's conserved; and so is symmetry. At any time in any space from any angle laws of physics stay the same.** Does physics see such changeless cosmos as a self-consistent, fine-tuned set of principles or are its invariant patterns of behaviour basically informed by chance?[34]

Rules are invisible but they precede a game. Rules are information, information is potential for behavioural patterns. Regulation thus precedes and guides the way a game is played spontaneously.

In this case we say that in the beginning was *NOT* chaos. Modern physics shows that mankind dwells in a finely-tuned universe. To be precisely fit for life it must have started in a way most orderly, specific, specially defined. *Fine-tuned by chance? If low probability together with specific definition indicate design, no chance!* **A universe fine-tuned regarding many parts might be construed as one of specific, irreducible complexity; and, if lacking any part it failed to work, of minimal functionality.**

No objects bear a number yet you number them and count. Numbers, symbols and mathematics aren't a physical but changeless, metaphysical reality. *Maths, while we're on the subject, is a real form of metaphysic.* Pythagoras, at least, believed that 'All is Numbers'. Could essential, physically-independent numbers really govern physical complexity? Einstein, Planck and Eddington believed that, once you dropped upon their key, you could *deduce* (*top-down*) the reasons for all natural laws; you might unlock the codes whose inmost mysteries reveal just how a stable universe is sparked; a feat of mind *par excellence*!

For others, including Roger Penrose, the Platonic world of mathematical forms is also real. He writes, '*There is a very remarkable depth, subtlety* and *mathematical fruitfulness in the concepts that lie latent within physical processes.*'

Could such immaterials help describe mind-matter frontiers whose formations are called archetypes? **Archetypes are also metaphysical and have no being but in mind. Maybe 'natural mathematics' is describing real, archetypal files.** *Perhaps archetypes compose the link between the corners of what Roger Penrose has referred to as a mysterious triangle of physics, maths and mind.* At any rate, their logic's fixed. Fixed forms of mind are memories; thus they are memories in universal mind.

Could *you* plan a cosmos better? Lucid physicists agree it looks 'as if' designed. **When it comes to astrophysics and cosmology stunning ingenuity seems to have coordinated chemistry and physics' natural laws, not least when it comes to bio-friendliness.**

Where, therefore, is the projector; what's the transcendent nature of projection's source? Natural Dialectic's 'holographic' edge is everywhere; it's 'super-posed' on physics, omnipresent but invisible because it's metaphysical. **It is the place where theme is turned to individual instance, principle is practised and where physic with its metaphysic meets.** Space,[35] time[36] and things are *within* universal mind. **Mind's archetypes project our world; archetypal memories, potential matter, are the essence of our physic; they inform, unchangingly, material being. Immaterial information holds the world, physical and biological, together. It is by archetype conserved.**

No doubt, in any case, we're here because the universe is as it is. That's no surprise. How came it so? Was its initial condition chance or not? The universal body is sharply defined by a precise set of over thirty interdependent settings. *Their values combine to generate a universal pin-code that was either preordained or at least intrinsic in primordial projection. Indeed, the dials are set for the sun, earth, you and me to an accuracy computed by Oxford mathematician Roger Penrose at 10^{10} to power 123!* **If true, that cuts chance completely out.** Erasure of coincidence. The probability of your bullet hitting a nail-head at the other end of cosmos first shot is vastly greater. Mathematicians consider odds longer than 'only' 10^{50} against to be zero. That is to say, there is statistically no chance whatsoever that cosmos and its dependent life are accidental. **The Penrose computation, if valid, indicates that odds against the observed, law-abiding universe appearing by chance are stupendously astronomical; and the facts appear to support his calculation**. The consequences of such statistical annihilation of chance.

Informative archetype (potential energy/ matter) is thought of as letters in a universal alphabet, as bits, bytes and routines of a computer program, an alphabetic cosmology. They would form the basis, as physics well observes, of cosmo-logic's script. Even better (since vibrations/ wavelengths correlate with forces and energies), notes whose harmonics compose the cosmic opera. Particles can be conceived as carriers of archetypal code; the behaviour of non-conscious substance is thus metaphysically controlled.

This perspective is, after a wave of materialism lasting over 150 years, neither fashionable nor the norm. However, why does an initial philosophical choice (such as, generally, science without thinking takes) make it impossible?

Now physics starts. We move now to consider the *top-level* or causative product of initial condition. We enter the kinetic phase of active energy.

In *top-down* terms the causative level of physical cosmos is metaphysical - archetype. From this template is projected, orderly, the kinetic, energetic level of quantum physics.

Materialism, however, lacks a reason for emergence into physicality. *It therefore relates the phenomenal start of a horizontal, age-related chain of cause and energetic process as what is nowadays conceived of as a 'big bang'.*

A Miraculous Projection

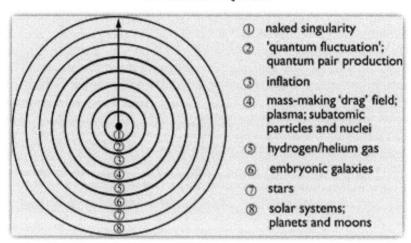

①	naked singularity
②	'quantum fluctuation'; quantum pair production
③	inflation
④	mass-making 'drag' field; plasma; subatomic particles and nuclei
⑤	hydrogen/helium gas
⑥	embryonic galaxies
⑦	stars
⑧	solar systems; planets and moons

Big bang is, in the standard horizontal sense, a magnificent story of creation with a physics-only slant. *Bottom-up*, we're told, a 'bang' *from absolutely nothing* did the trick. *Look at your thumbnail.* Imagine once again that, for no reason, force for a universe jumped out from nowhere (at least a point far smaller than the nail). You are but a microscopic fraction of galactic clusters that, a speculator called *Ad Hoc* is claiming seriously, erupted from a microscopic, smaller-than-a-pinhead fraction of that thumb - unless he paradoxically means the primal dot in nothing's everywhere!

	grade	time
sat	*pre-physical latency*	-
raj	*quantum micro-level*	*Planck/ quantum era of high energy and quick time*
tam	*classical macro-level*	*era of low energy/ bulk aggregates and slow time*

Yet following the drift of Lady Luck allow the labours of her empty womb. Set your stopwatch. Press! After naked speculation of a singularity assume, from an abyss of nothingness, things miraculously start to rise.

The greatest trick, the luckiest break of all time! Barren nothingness has, in some way, been fertilised! From such 'electric' touch an ovoid fluctuation in the timeless, spaceless emptiness uncurled; this minuscule and trembling instability is due to be Creation's Seed. In only billionths of a second swirling particles emerge from the Great Generator's heat.

Next, bang on bang on time, a second, bigger whoosh occurs Because the first exertion would have been in danger of collapsing back inside itself you need a second miracle to free it up and drive the show onto the road. 'Cool expansion' grew until the universe was human-head-large; then 'hot inflation' had to blast things to the skies. This *Ad Hoc* inspiration was devised to blow the first bang up a trillion times in billionths of a second. It exploded much, much faster than the speed of light then, like the first one, braked just right. Thus an assumptive boost of cosmic after-burner, inflation with inflaton fields, has ironed out all irregularities and, most unusual for a double bomb, set things bang on track for how they are today. Bang! Boom! Simplicity itself. **It might be, as we observe from 'inside' physicality, start's smoking gun.** Therefore what headache, even to a physicist, could such a grand yet simple vision of creation present? Sadly, pass the paracetamol.

Labours of an Empty Womb! Did it all begin by chance or not? There's no business like show business! What force of creativity invigorated and invigorates a latent void's virginity? Are 'things' themselves the fruit?

Something for nothing (or from nothing) is, for physics, basic nonsense. Thus, for starters, let's realise that first appearance of the physical dimension and subsequent description is no more an explanation of its zero-point than explanation of your development from its own egg. *In both cases what kind of 'nonsense' pre-initialised? On what metaphysical priority does each case depend?*

Must mindless Mother Nature have conceived herself this way? Is natural law, the 'fix' of physics, child of Lady Luck? Natural reasons have no reason; nor do material effects have *a priori* cause. *Certainly there's no intention that substantiates cosmology.* Freeze teleology right out of every frame - you can speculate about the universe but only, trained on *bottom-up* perspective, from *inside* a totally materialistic order of the game! While sharpening lines round any passing cloud of thought no speculation on an immaterial factor, a metaphysical, informative dimension is allowed. And so, of nothing-physical, you ain't seen nothing yet.

Yet unseen but fecund void, called archetype, is where *top-down* causation metaphysically, that is, pre-physically begins. *In Natural Dialectic's holistic, multi-layered spectral view, exterior appearance stems from subtle inner bands; creation issues, layered, from within.* **Cosmos not chaos sets initial conditions up. Called potential matter,**

archetypal files are seen as hard a metaphysical reality as, say, particles are physical realities. Plan, principle, directive. Such potential comprises program(s) naturally stored in cosmic memory - simple in terms of inanimate physical 'law' (of particles and forces), complex in terms of animate structure/ function/ behaviour.

Cosmos not Chaos

tam/ raj	*Sat*
existence	*Essence*
exhausted/ kinetic phases	*Potential Phase*
variation	*Invariance/Permanence*
development	*Seed*
practice/ actuality	*Principle/ Law*
subsequent order	*Archetype*
elaboration	*Plan*
change/ inconstancy	*Stability/ Constancy*
↓ *tam*	*raj* ↑
passive state/ object	*active energy/ force*
exhausted phase	*kinetic phase*
informed/ ruled	*informant/ ruler*
contingent variation	*basic themes*
external 'fall-out'/ outcome	*internal order*
crystallisation/ bulk shape	*energetic pattern*
chance	*necessity*
accident	*design*
apparent chaos	*cosmos*

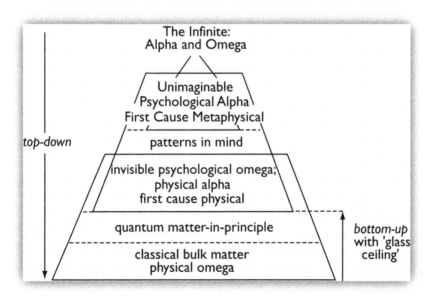

79

The previous and the next couple of illustrations **relate the dialectical order of cosmic emergence.**

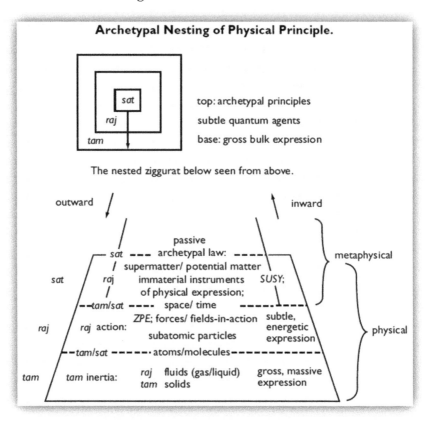

In this way sub-atomic elements, forces and atoms of physics and chemistry can be construed as basic elements of code. Such code would dictate, through the agents that a study of phenomena elucidates, the way things naturally turn out. Quantum particles can be analogised to alphabetic notes/ marks/ letters of a cosmic language.

physical matter	*Potential Matter*
action	*Physical Absence/ Void*
program played	*Archetypal Program*
lower orders	*Upper Syntactical Level*
↓ *matter-in-practice*	*matter-in-principle* ↑
gross/ extrinsic constraint	*subtle/ intrinsic motions*
classical bulk	*quantum flux*

This is the nature of our cosmic egg. All eggs are codified in order to develop in coherent ways.

> *'Egg' is an apt metaphor for outworking from within; but development from an egg, far from being a random process, is preconditioned and involves precise codification.* **And codification, demanding forethought, is a product of mind.**

Two stacks dialectically express nature's bauplan or blueprint; and help to show how archetypal nucleus gives rise to material-expression-in-principle (the quantum substrate) and in practice (aggregates).

Looked at this way cosmos is a Grand, Dynamic and Encoded Text.

tam/ raj	*Sat*
yang/ yin	*Tao*
down/ up	*Axis*
oscillation	*Point of Balance*
cycle	*Centre*
↓ *down*	*up* ↑

From here, therefore, you can relate the character of basic players to cosmic fundamentals. **Nothing is added to the facts of physics except unification within an overarching and holistic plan.** Yet such *GUT* is essential to full and inclusive understanding of the world's inception.

tam/ raj	*Sat*
expression	*Idea/ Purpose*
↓ *tam*	*raj* ↑
passive element	*active element*
informed	*informant*
physical elaboration	*conceptual elaboration*
created outcome	*creativity*
output	*input*
end-product	*development inc. manufacture*
hardware	*software*
machine	*program*

Textbook physics casts its order in somewhat different ways.

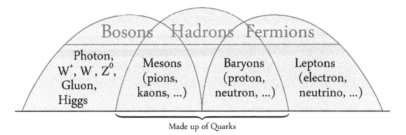

This triplex is developed by a standard model:

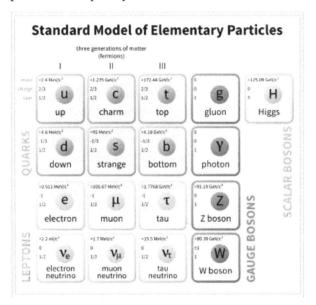

And you can tabulate elementary particles in terms of matter and force:

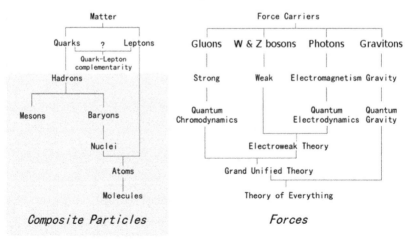

Physics is, of course, the study of energetic interactions of precise, pre-set and wholly automatic kind; and, put simply from a dialectical point of view, chemistry observes the push-and-pull of polar charge.

But what, one asks, *is* the cosmic glue called charge? What is an electron but locked vibration, standing waves; what is an atom but a tiny oscillator? From non-local, omnipresent archetype are issued the automatically, mindless behaviours we call rules - specific, rigid laws of nature. These behaviours are harmonic oscillations. They manifest locally and are known to physics as the quantum world of particles and forces; they are the minuscule and energetic base from which, as icebergs in a sea of energy, gross sensible matter is formed. Natural Dialectic calls this level of creation, the quantum phase, subtle matter or **matter-in-principle**.

Physics deals in detail with *quantum and atomic physics* whose simple, primary agents give rise, in congregation, to subsequent complexity. You might think of quanta as the 'inner nature' of gross matter from whose subtle basis our gross world is formed. These subtle quanta precede gross aggregations. Physics delves from gross to subtle. Whence, however, first arose primal simplicity and interactivity? What, moreover, *sustains* the substance of a quantum like a quark, proton or electron through billennia? No doubt, *Karl Popper dubbed quantum physics 'the transcendence of materialism'* but for *Natural Dialectic this phrase applies to the metaphysical element that we just dealt with called potential matter/ archetype.* Archetype, if anywhere, precedes and is *within* the quantum state. It is matter's pre-active 'field', its latency before quantum 'becoming' and, in atom, molecule and gross forms, what has become material phenomena.

The problem is, as usual, with prepositions. Your mind, you might agree, is *natural:* but is it 'at' or 'in' or 'of' your brain? Is its maker's mind 'in', 'behind' or what regarding even the simple seat you're using? Where is an e-m wave with respect to a radio - within or all around it? Is universal mind within each particle of physics' zone or does it, paradoxically, embrace the whole material universe 'within' itself?

$$\vec{\nabla} \cdot \vec{D} = \rho$$
$$\vec{\nabla} \cdot \vec{B} = 0$$
$$\vec{\nabla} \times \vec{H} = \vec{j} + \frac{\partial \vec{D}}{\partial t}$$
$$\vec{\nabla} \times \vec{E} = -\frac{\partial \vec{B}}{\partial t}$$

Light, whose man, Clerk Maxwell, helped usher in our technological age, is the least massive form of energy. *Its particle, the photon, has no mass. Nearly nothing, it is in practice physic's holy ghost.* This ghost is absolute regarding its velocity, that is, its relativity; and, for natural dialectic, closest to the metaphysic of potential matter. We know that its 'pure' energy is transmitted by vibration, that is, in waves. We also know that light can carry a great deal of information and, if focused, laser power. Could information have been first injected into cosmos on the back of light?

At any rate, complementing chargeless, radiant light is charge. Each particle (except light) theoretically has a polar opposite, an anti-particle identical except for opposite charge. Charge (polarity) is fundamental to the way cosmos works but what *is* it? What *is* an electron? Why that exact strength of charge or that particular carrier as opposed to something else? Why, when they hit each other, do particle and anti-particle annihilate into gamma-rays (light)? Why is charge a set unit of power except, possibly, in instances where such parity is violated?

Of **basic, polar opposites**, what *is* their cosmic glue? What is a communicative *electron* but locked vibration, standing waves? What, on the other hand, is the reason for a complementary, isolating and contractive force initially expressed as strong nuclear in quark or proton? What, thirdly, is the balanced structure called an *atom* but a tiny oscillator?

We can also ask, in dialectical mode, how electrons and quark/protons (that with neutrinos and neutrons form the basis of our known 'low-energy' universe) so enduring, to the point some calculate the total number of them?

Such minuscule behaviours as we've described *are* harmonic oscillations (see Lecture 3). They manifest locally and are known to physics as the quantum world of particles and forces. In other words, from non-local, omnipresent archetype are issued the automatically, mindless behaviours we call rules - specific, rigid laws of nature. They compose the minuscule and energetic base from which, as icebergs in a sea of energy, gross sensible matter is formed. Natural Dialectic calls this level of creation, the quantum phase, subtle matter or **matter-in-principle**.

You can look through pink or green-tinted lenses; you see the same fact in a different way. The holistic, dialectical view does not change the shape of facts but shifts perspective. It shifts their context to another angle, to a larger-than-materialistic fit. How does dualistic/ triplex Natural Dialectic use its cosmic fundamentals of (*sat*) equilibrium with (*raj*) levity and (*tam*) gravity to describe a 'three-dimensional' fit? Was it such cosmo-logic or its lack that started everything? Here is a brief introduction to the pattern as expressed by a few illustrations.

You could begin by setting out a triplex starter.

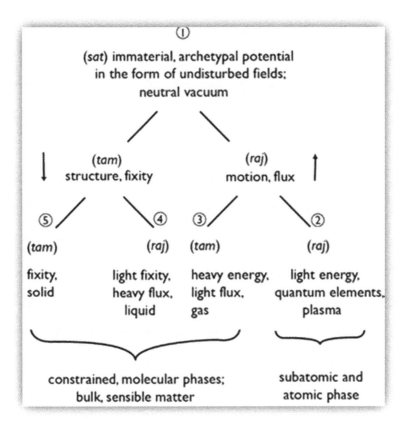

Perhaps science seeks, using somewhat esoteric terminology, for more mathematical complexity than this description offers. Is there, however, deep *simplicity* behind phenomena (such that, for example, you might dialectically predict a space-linked force of levity, anti-gravity or lambda)?

Communicator, messenger or force particles are called bosons; massive particles, the basis of matter, are called fermions. Decay from high to low energy drops through three generations of material constituent (fermion). The present, third and (*tam*) lowest energy generation comprises our leptons (neutrino and electron) and quarks (proton and neutron from so-called 'up' and 'down' quarks).

You rightly ask what activates creation's grid. What revs absolutely nothing up? For materialism this 'something that is nothing' even precedes the emptiness of space. How can nothing rev up nothing to a universe? If this is nonsense could eternal matter do the trick? Yet steady-state theory is for several reasons a rejected entity.[37] So, left with no alternative, perhaps eternal, multiversal energy has slightly stolen physic from its inescapable abyss.[38]

On the hand, holism holds that physical nothingness, in the form of archetypal memory in universal mind, is activated by a 'lower voltage' of the Archetypal Cosmic Dynamo.[39] This unconscious grade of 'electricity' at once stimulates and sustains vibration that, in vacuum, shows as force and particle. It is, in other words, the 'wind' that activates the instrument of archetype.

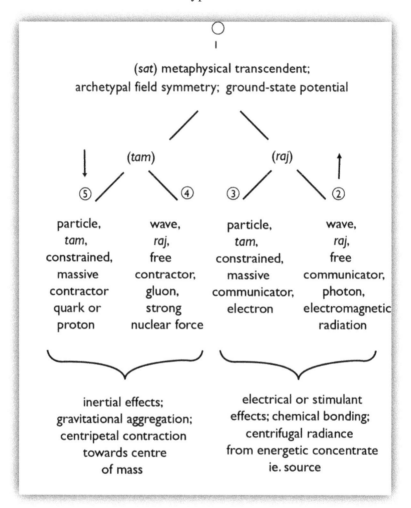

As polarity arises from the division of neutrality, duality from the division of unity and motion from a loss of poise, so you might expect a principle of equilibrium to source such symmetries (for example, gauge and particle symmetries or the numerical 1:1 electron/ proton charge ratio symmetry) as underlie the fabric of space. We've seen that such symmetrical wholeness would, according to Natural Dialectic, be metaphysical. Actual physical particles and their bulk phenomena would not derive from or be different appearances of a single kind of underlying particle (say, a hypothetical preon); nor would they

86

transmute into each other (neither protons nor electrons, for example, are seen to decay). Instead they would represent the activation of an intrinsically-polarised archetypal latency; they would represent disturbance of initial balance. An example is the quantum pair production of particles (say, an electron and positron) from the vacuum; and their mutual annihilation into neutral light. Broken symmetry leads, in other words, to polarity, duality, waves, particles, sub-principles or laws of nature and a perpetually unbalanced, changeful, motile creation. **In the dialectical metaphor of scale norms (or laws) are axial; nature can be seen as a swinging, vibratory attempt to forever regain a perfect balance.**

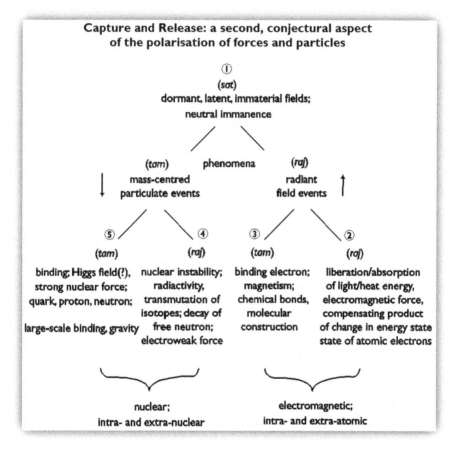

In this model of essential balance cosmic motion is seen in terms of equation or equilibration. In this way the universe can be seen as a vast homeostatic process wherein forces are the agents of negative feedback; they are balancers, always seeking equilibrium. They are the expressions of a natural attempt, after Initial Disturbance (for example a big bang), to regain norms, equilibrate action (with equal and opposite reaction) and 'keep account'. For psychological, 'upwardly mobile' forces this equilibrium is one of poise - at maximum the Super-state Poise of Enlightenment; but for 'downwardly mobile' physical forces it is one of inertial state, bonded aggregation and exhaustion - whose maximum is found in the sub-states of Absolute Zero (0°K) or a black hole singularity.

In this view both polar forces and particles are themselves agents of (*sat*) regulation. Metaphysical preordination of particular expressions of the cosmic fundamentals (e.g. photon, electron and proton) is the (*sat*) axial principle around which the (*raj/ tam*) scales of cosmic motion swing. In this case there would exist more than a couple of dialectically polar sides. *There would be three sides to the single story of cosmic homeostasis - two physical, one metaphysical. Actual matter would take shape from metaphysical potential. Is, therefore, the origin of natural law just accidental? Or does it follow archetypal rationale?*

Let's re-simplify using the sort of triplex stack we met in Lecture 1:

↓	*tam*	*Sat*	*raj* ↑
	down/ descent	*Pivot*	*up/ ascent*
	gravity	**Balance**	**levity**

It's clear that dialectical polarity is not confined to electrical charge. It applies to cosmic swing. This oscillates according, within (*sat*) regulatory balance, to a couple of vectors. One, (*raj*) negentropy, is action's wake-up call; it corresponds to dynamic stimulus and/or an increase in configurative complexity. You might expect its action to predominate at starts. The other, (*tam*) entropy, is action's fall towards sleep; it corresponds to inertial or disintegrative tendencies towards exhaustion, differentiation and disorder in the sense of randomness. You might expect it's how you'd find things fall apart while rolling towards their end-game.

Dual aspects of the vectors need to be distinctly underlined. In dialectical terms what exhausts or fixes is a (↓) gravitational effect; the opposite is (↑) levitatory. In dialectical as well as scientific terms (↓) entropy and (↑) negentropy each have two aspects - informative and energetic or, if you like, configurative and thermal. When purpose is excluded then, as nature flows, these two show inverse effects. If temperature is turned up then thermodynamic negentropy shows configurative entropy (the 'randomised disorder' of gas particles); if heat is lost thermodynamic entropy shows configurative negentropy (in the sense of atomic crystalline or solid order). Yet overall, as in the case of a steam locomotive, hot gas will cool and 'lose capacity to work'. What is a crystal but assumption of the slackest shape, least tense configuration and expression of inertial equilibrium? Without a purpose or without a fire the world is tumbling down its hill. Sooner or later, underneath their start, all physical creations fall apart, go up in smoke or fade away.

tam/ raj	Sat
individual expression	Potential
local	Global
specific	General
action	Preconditional Poise
gravitational/ levitational	**_Archetype_**
negative/ positive	Neutrality
end structure/ change	Original Balance
flux/ change	Stability
↓ tam	raj ↑
negative	positive
end/ negative dominant	start/ positive dominant
lock	activate
materialise	dematerialise
less free energy	more free energy
entropy	stimulant negentropy
collisions/ pressures	unobstructed movement
object	event
contractive	**_radiant_**
flat/ straight line	wave/ vibration
strong nuclear	electromagnetic
point force/ localiser	expander/ delocaliser
strengthen with distance	fade with distance
create locality	make space/ transmit
isolate	communicate
proton (microscopic)	electron (microscopic)
mass (macroscopic)	shine (macroscopic)
gravity	**_levity_**
mass-to-mass attraction	buoyancy of space
mass dominant	extension dominant
discharge/ end structure	flux/ reactive change
sub-state neutrality	interaction
atomic structure (micro)	chemical reaction (micro)
micro-stability of atom	interactive (electronic) glue
aggregate (macro)	large-scale dynamic flux
macro-stability of solid	classical motions

There are many stacks that you can write connecting aspects of duality and trinity. Let a triplet more suffice for now.

consequences	First Principle
expression	Potential
local, physical types	Singularity
corresponding phenomena	Archetype
thing	Nothing Physical
↓ informed	informant ↑
outcome	principle-in-action
anti-principle	principle
apart	together/ co-operant
unit 'self'	connector
specific/ detail	general sketch/ outline
bulk events	atomic/ sub-atomic events
classical level	quantum level

Physic emerges just below the start. Each from its archetypal *noumenon* material *phenomena* of space, time, particle and force appear. These ghosts are the players that compose our stage.

Ghosts in the Field	
relativity	Absolute
division	All-in-Oneness
polarity	Neutrality
fermion	Boson
hadron/ lepton	Photon
↓ minus	plus ↑
hadron	lepton
quark/ proton	electron
contractive fields	electromagnetic field
massive/ bulky	lightweight/ lifting
gravity	levity
drag together	stimulate/ drive apart
confinement	liberation
binding/ capture	freedom
fixity	mobility
particle form predominant	waveform predominant
locked energy	connective energy
isolator	communicant

This, last in line, is both beginning and the end of matter.

existence	Essence
created events	No Thing
cycles	Centrality/ Hub
outcome/ action	Pre-condition
expression	Pre-active Potential
polarity	Neutrality
sink/ process	Archetypal Source
alphabet	Alpha = Omega
↓ informed	informant ↑
passive energy	active energy
classical (bulk) phase	quantum phase
entropy	stimulant
end/ death	process/ lifetime
exhausted void	action
impotent neutrality	drive
sink/ periphery	flow
omega	alphabet

To complete the cycle, though, we need to turn to the creation's edge, the world's periphery.

(Tam) Passive Energy

Passive energy is known as the classical phase of bulk matter. Such matter-in-practice is considered to be in locked, exhausted phase. Its 'bonds' include molecular formations, plasmas, gases, liquids, solids and, of course, all study related thereto. It represents projection's furthest radius from Source. The subjects of physics and chemistry deal exhaustively with this level of creation.

For this section *let the final word rest with Max Planck*. Planck not only first read, recognised and published (in the *German Annals of Physics*) the start of Einstein's revolution, the latter's Special Theory of Relativity. He also pioneered quantum theory, the second pillar of modern physics whose study is sub-microscopic, sub-atomic phenomena - the *matter-in-principle* of Natural Dialectic. Actually, his foresight may have ushered in the next revolution in human understanding which has already begun to focus on the primacy of the 'unscientific' informative co-principal as opposed to its energetic coordinate. He asserted that the discovery of truth can only be secured by a determined step into the realm of metaphysics and is quoted as saying at a lecture in Florence (1944):

"As a man who has devoted his whole life to the most clear-headed science, to the study of matter, I can tell you as the result of my research about the atoms, this much: *there is no matter as such.* All matter originates and exists only by virtue of a force which brings the particles of an atom to vibration and holds this most minute solar system of the atom together...We must assume behind this force the existence of a conscious and intelligent Mind. This Mind is the matrix of all matter."

Indeed (The Observer 25-1-31), "I regard consciousness as fundamental. I regard matter as a derivative of consciousness."

Perhaps Max Planck would have appreciated Natural Dialectic.

Summary: In Natural Dialectic's spectral view, exterior appearance stems from subtle inner bands; creation issues, layered, from within. From nuclear programs issue flowers and children. So do rocks and clouds and stars. As we saw in the last lesson, a code is an agreed set of symbols arranged to format information. Such **upper linguistic/ codified level** involves particles (say, letters of an alphabet); forces that regulate their conjunction (punctuation); grammar (say, the elements of a language such as atomic noun, verbal motions and so on); and syntax. Syntax is the convention or legal framework within which symbols are ordered; its law naturally determines those structures allowed and those not. Thus the *upper syntactic level* acts as a filter through which order is communicated to and from the *lower (environmental, statistical or quantitative) level* of data items.

Another view of creation is, we said, 'organic'. *'Egg' is an apt metaphor for outworking from within.* To repeat, the cosmos is projected from a seminal conception and its consequent, logical ramifications; conceptual development is stored, like the blueprint for any construction, in memory; the blueprint's physical code is represented by an 'alphabet' of simple particles with a forceful 'grammar'; and a starry universe is, finally, the outermost fulfilment of A Plan.

In this case physical nature is informed by an alphabet of sub-atomic letters whose interactions (by four forces) build into a large vocabulary; they sing the saga, compose the world's text not on air or paper but the blank of space; they use the particles of physic's language; they represent a cosmo-logical code.

Lecture 5:   Biology

Before the 1950's biology was a matter of 'outward' observation, copying, classification and, for about 30 years, some 'elementary' biochemistry (of vitamins, proteins and a suspicion, no more, that a substance called nuclcic acid was a central component of all living things.

After 1953, however, things really took off. The structure, operation and 4-letter alphabet of a superb computer language was discovered. *DNA* is to biological form what computer chip is to an AI robot - except far superior in operation in that, like an entirely automated factory, it can cause the reproduction, growth, maintenance and repair of itself and the form it inhabits.[40] *DNA* codes, through an intermediate called *RNA*, for protein. Above are ranged discoverers Francis Crick, James Watson, Rosalind Franklin, Maurice Wilkins and 'runner-up', Linus Pauling.

A founding father of molecular biology was Malcolm Dixon. He was followed by Fred Sanger and Max Perutz who, again here in Cambridge, led the field in sequencing amino acids that compose a protein. In the same lab at about the same time Sydney Brenner realised that *DNA* was a language. Actually it is a digital quaternary (as opposed to binary) code; and, on top of that a double code. It may even be that, as well as epigenetic considerations, some sections of code may read productively either way; such palindromic sense would require a very high level of ingenuity. However this may be, now *DNA* itself can be rapidly sequenced. Soon we shall have a library that includes the texts of all living organisms.

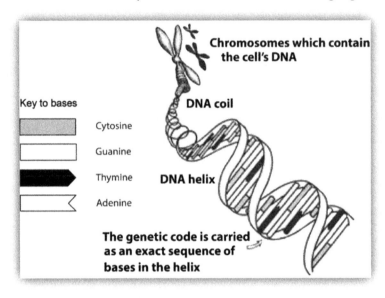

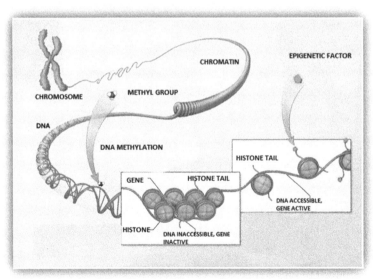

Of course, epigenetics - codes governing codes - amounts to extra informative dimensions.[41] Hierarchically arranged, its mechanisms build a towering, high-grade data structure. *A responsive genome incorporates multiple, overlapping and interactive code in a mode the IT profession calls data compression.* **How can first-class data compression arise without a**

prior informant? Ask any software engineer. *Why should DNA and its controllers, packing information just as tight as any hard drive, be different?* So the basic bio-issue is, unrelentingly, the source of code.

Naturalism is tenaciously defended. The metaphysic of its mind-set is sacrosanct. And while physical study of biological form and function is entirely reasonable, in order to satisfy naturalistic methodology's need for materialistic answers alone, a story of how life's forms appeared on earth is needed. Hence Darwinism is cemented into scientific world-walls; trespass on that space is charged with secular anathema. Do you, however, recall that information is an immaterial factor and, as we well know, it forms the very basis of biology? **In this case the basis of biology were metaphysical.**

Biologist Theodosius Dobhzhansky is famous for coining a popular mantra: 'nothing makes sense in biology except in the light of evolution'. **But the actual, iconoclastic fact is: 'nothing makes sense in biology except in the light of information.'** You may drag evolution in on information's tail but it is simply a fashionable word used in biology to mean several different things. *In reality, codes and signals run the bio-show.* Not evolution but information is the basis of biology. To recapitulate: it is not, trivially, an origin of species but, fundamentally, the origin of information that is at issue.

When will materialistic science come to grasp the implications of this fact? How, within materialism's paradigm, can it be reasonably addressed? The only innovator matter has are its oblivious behaviours interacting due to circumstantial chance. *No plan, no program, utter aimlessness.*

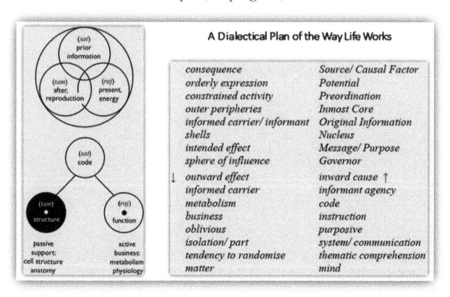

A Dialectical Plan of the Way Life Works	
consequence	*Source/ Causal Factor*
orderly expression	*Potential*
constrained activity	*Preordination*
outer peripheries	*Inmost Core*
informed carrier/ informant	*Original Information*
shells	*Nucleus*
intended effect	*Message/ Purpose*
sphere of influence	*Governor*
↓ *outward effect*	*inward cause* ↑
informed carrier	*informant agency*
metabolism	*code*
business	*instruction*
oblivious	*purposive*
isolation/ part	*system/ communication*
tendency to randomise	*thematic comprehension*
matter	*mind*

Translated to biology such innovation is perceived as sexually-

95

produced variety and genetic mutation. Is this pair sufficient to generate the texts of life on earth? We shall see.

First, however, in order to sensibly discuss an inclusive, holistic alternative, let's suggest a brief but broadened theory of biology. We've already looked at archetypal memory, first cause of all things physical. How, amid the specific, functional and fully codified complexity that is organic form, is this expressed? Here we see and follow the triplex, dialectical order of expression.

consequence	*Source/ Causal Factor*
orderly expression	*Potential*
constrained activity	*Preordination*
outer peripheries	*Inmost Core*
informed carrier/ informant	*Original Information*
shells	*Nucleus*
intended effect	*Message/ Purpose*
sphere of influence	*Governor*
↓ *outward effect*	*inward cause* ↑
informed carrier	*informant agency*
metabolism	*code*
business	*instruction*
oblivious	*purposive*
isolation/ part	*system/ communication*
tendency to randomise	*thematic comprehension*
matter	*mind*

Information precedes.[42] It is prior and anticipates. Information is the potential and *sine qua non* for the action that consequently issues orderly.

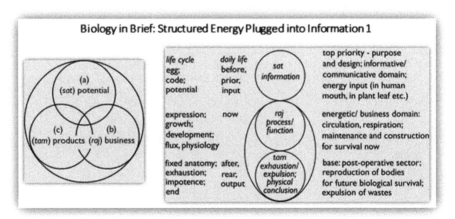

Information comes, as previously explained in Lecture 2, in two forms - active/ conscious and passive/ unconscious. The unconscious, informed case shows as archetype and soft, bio-logical machinery.

96

Biology in brief: structured energy plugged into information. The **unconscious, informed case** also involves both psychological and biological components. These include archetype (instinct and morphogene); reflex balance, called homeostasis, by way of nervous, hormonal and other systems; cybernetic metabolism controlled by preordination in the form of genetic code carried chemically by *DNA*; and muscular organs of action and response. We discussed archetype in the sections on psychology and physics. In regards to the latter, the assertion that particles and forces constitute an alphabet, punctuation and grammar of a fine-tuned text can be met with a shrug; really, it was accidentally formed. *With biology's complex computer code and operations this view changes dramatically.* If cosmos is fine-tuned why not biology as well, with great but very accurate complexity? Suffice, at this point, to reiterate that if I asked you for physical proof of mind in the chair you are using, you would rightly dismiss the question. Yet even a simple chair has the mind of its designer definitely there. Similarly, mankind may eventually come to understand every last quantum of soft, codified bio-machines but will this be all? The most important element, the source of information needed to construct them in the first place will have been ignored. Such explanation is, as Polyani noted, in no way complete.

Science has such poor explanations for 'purpose', innovation and detailed, specific construction of mechanisms that surely a valid alternative is archetype?

Life's process is one of dynamic equilibrium. Equilibration. Its goal is balance in accord with pre-set norms. Metabolism, being totally information-dependent, works with reference to precise, incoming

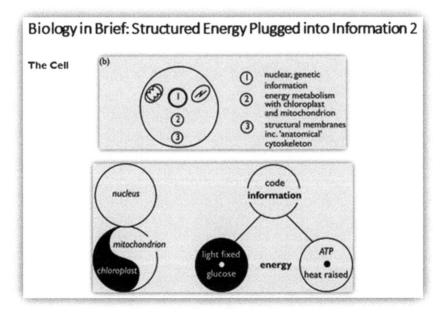

messages and equally precise genetic response. Such fixed response is indexed, switched and flexibly monitored by non-protein-coding and epigenetic factors; and also by interconnected nervous and hormonal systems. On the other, psychological hand, a flux of desires creates a moving set of targets whose equilibration (or neutralisation) is reached in satisfaction.

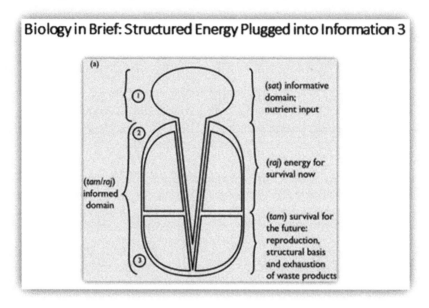

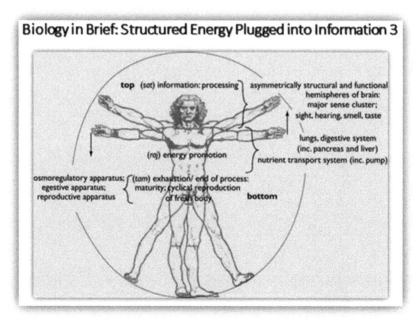

Life is, in this way, an incarnate flux of order due to information. **In short, organisms are information incarnate**.

Energy provides for survival now. It involves, metabolically, photosynthesis and respiration. It promotes cell biochemistry, trans-membrane voltages, physiological processes and, on the large-scale, (nervous) sensation and (muscular) motion. The character of all function is energetic.

Structure, whose character is solidity, represents the outermost, fixed (or flexibly fixed) realisation of shape. Its base domain is energy's container, a fixed expression of internal, orderly flux. In other words, a 'phenotype' is a peripheral aspect whose body both reflects and fixes the shape of inward information and energy. The end-product of structural development, maturity, is reproductive. The cycle starts again.

These two slides plug information into a larger body. For Natural Dialectic a cell is primarily information, only secondarily biochemical. And, as we see, the structure of your own body is an expressions reflecting the cosmic fundamentals.

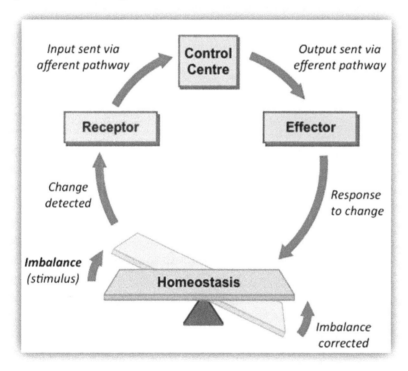

We have nuclear super-computing but, in the energy department, the central executive is homeostasis.[43] It is based upon the geometry of periodic action, a circle and its extension, vibration. Vibration round a norm is at the root of biological equilibration, its dynamic equilibrium, its energetic stability or superb balancing act. The controls are set for balance; these controls are always, in principle, triplex. They involve mechanisms of sensor, processor/ controller and effector. *All homeostasis, in any cell, must involve these three deliberate components.*

Why, in the face of physical entropy, should equilibration be the goal? Why, in spite of nature's fall towards exhaustion, breakdown, garbled chance events and death, should a tight-rope for survival be so evidentially codified? Indeed, why should a group of atoms ever 'want' to survive? Is not the whole cybernetic business of homeostasis for stability of operation? **Cybernetics intrinsically involves anticipation, purpose and a goal.** Biological energy is, through documented process and specific agents, strictly guided; it is, in a word, completely informed.

Code anticipates. Signals communicate. Information operates with feedbacks which ensure dynamic systems 'stay in line'. Therefore, to find the causes of 'downstream' effects you have to travel to their 'upstream' cause. *Such cause is not by natural selection. Nor is it mutant lesion of a gene or chromosome.*

Hierarchical information-structures are conceptual. In other words, visible characteristics called phenotype depend on molecular action that, in turn, depends on genotypic and other pre-programmed information. <u>*Life is so obviously purposeful in character that genetic chemicals are ascribed all kinds of animistic properties.*</u> They 'compete', 'organise', 'express', 'program', 'adapt', 'select', 'create form', 'engage in evolutionary arms races' and even, lyrically, 'aspire to immortality'.

The origin of hierarchical order, cyclical flows of information and integrated function is always purpose. **The origin of purpose and its attribute, meaning, is mind.** *We might, therefore, reasonably infer that the basis of meaningfully informative, functional and structural biological hierarchies is mind.*

After informative, biology's second set of hierarchies is energetic.[44]

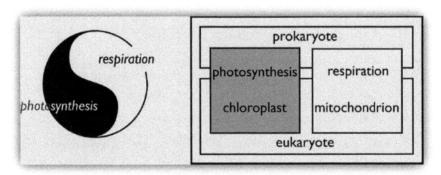

First in line stands the photosynthetic/ respiratory cycle with, in the case of eukaryotes, associated organelles.

Nutrients are 'frozen' out; a rain of nectar and a snow of sugar crystals is precipitated from the leaves of plants. *Fixed sunlight* is next joined by energetic gas; oxygen fuels, like a bellows, respiration. This regenerates a bonded 'safety-match', a 'battery' called *ATP*. From this universal currency of 'grounded light' is struck a flame to drive metabolism and the chemistry of everybody's body. Energy metabolism is a key operation whose two aspects, *anabolic photosynthesis and catabolic respiration*, underwrite the 'buzz' of energy we know as life.

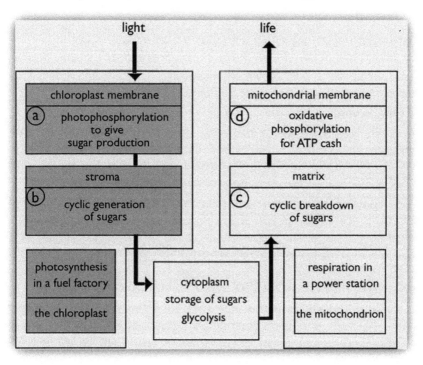

tam/ raj	sat
complex	simple
practices	principle
effects	cause
conversions/ transductions	light
↓ tam/ yin	raj/ yang ↑
anabolism/ small-to-large	catabolism/ large-to-small
light capture/ absorption	heat release/ emission
storage/ fixity	action/ flux
bond made	bond broken
light-to-sugar	thermal unit (ATP)
'earthing'/ materialisation	'quickening'/ dissolution
photosynthesis	respiration
fuel factory	power station
chloroplast	mitochondrion

Whence came the information for all these essential systems? *Bottom-up*, the problem's to convince yourself that, granted a vastly complex starting-point, the computation of cell chemistry haphazardly, mindlessly 'improved itself'. What mindful systems flow-chart could you build to demonstrate this possibility?

Top-down, programmers know that, from a main routine, switches branch to sub-routines and, when a sub-routine is done the process cycles back to start again. **These routines are modules**. This conceptual character of algorithm, this repetitious use of switches and blocks of modular code is just what coded bio-systems show. It is how nature's life-forms, full of reason, always work.

One would, therefore, predict research will more and more reveal signs of bio-logic to the point that, in 'live' computing, such complex permutations, integrated combinations and hierarchical sets of regulation will further squeeze then nullify the notion that celled systems ever cropped up accidentally, that is, evolved.

A computer is a mind machine.[45] *On this basis it is established that the cybernetic operations of a cell, an object as thoroughly material as a computer, superbly meet the criteria that pass it as a mind machine.* A cell is a mind machine. **Its instruction manual is a program written up as 'genome'; and its systems hold semantic meaning as modules or entities of bio-form like, for example, you.**

Various tailored sub-routines are called from a Main Routine. Suites of such modules combine as permutations around which different bodies are, under the coordination of such an Archetypal Master Routine, expressed.

Molecular biology's great strides are also heading straight towards the notion of an archetype. A similarity of genetic instructions is found to extend across a great variety of forms. For example, a high- level developmental complex of modules (called 'homeotic genes') may code for outline body-plan in very different organisms such as human, duck or fly. Or it may call subroutines for the construction of completely different kinds of eye - a normal gene from a mouse can replace its mutant counterpart in the fly; now an eye, a fly's eye not a mouse's, is produced. In other words, such genes act as 'go-to' switches in an archetypal program of development.

Or examine just one of thousands of molecular machines.

The next picture shows one component of respiration, a 'dynamo' called ATPase. It is used to 'burn' the sugars created in an organelle found in plants and phytoplankton called a chloroplast. Energy metabolism is composed of a duo, build-up and breakdown of specific molecules. We now know that its complex engineering includes, at the initial stage, quantum biology as regards the resonant energy transfer of electron excitation in a coherent, most efficient way

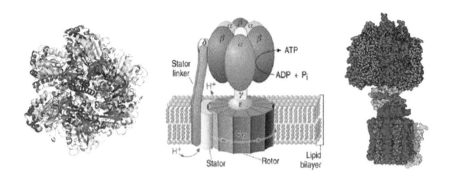

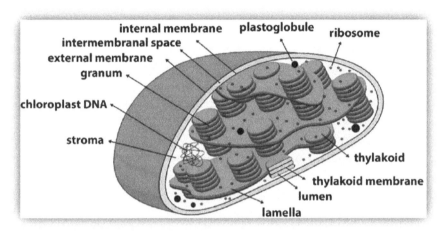

One could add much more but there, in the question of the natural source of information, rests the downfall of Darwinian scholarship and all its institutions. **The last 10 minutes have in fact demolished it.** No wonder there arise acutely-felt objections!

So why, if we know code is purposeful and issues from mind do we have to resort to such irrationality as chance construction (by mutation) of life's books. The answer, of course, solely to maintain materialism's grip. Question the theory of evolution and see no less than academic rage brew up.

Evolution? Check the Glossary. There are today *four* main usages of this word; each 'loading' derives from the original Latin, 'evolvere', meaning to unroll, disentangle or disclose.

So to a critique of the biological theory of evolution in four parts: natural selection, mutation, sex and development.

A half-truth is the most difficult to unravel. Elements of Darwin's theory are, of course, agreed by everyone. Everyone agrees that variation-on-theme continually occurs. Such variation is the result of sexual reproduction (which contains in own in-built lottery called meiosis) and, deleteriously, by genetic mutation. *Top-down*, **it can also the product of adaptive potential.**

Perspectives on Three Central Tenets of neo-Darwinism - a Tabulation

		✓ true ✗ false	Bottom-up	Top-down
①	Abiogenesis		✓	✗
②	Variation (microevolution) by mutation and natural selection		✓	✓
③	Transformism (macroevolution) by mutation, natural selection or any other means		✓	✗

Except for cases of bacterial resistance to antiobiotic chemicals it is hard to nail down a single mutation that does not detract from an organisms coherence. Do random errors in a script improve it?

Remember what was said at the start? **Darwin asked the wrong question.** *With what we now know the question does not concern an origin of species but, fundamentally, the origin of information.* Nevertheless, let's take a look at speciation.[46]

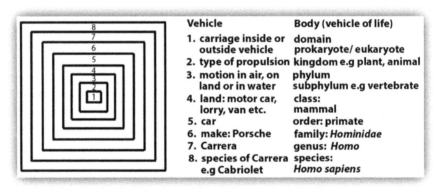

Vehicle	Body (vehicle of life)
1. carriage inside or outside vehicle	domain prokaryote/ eukaryote
2. type of propulsion	kingdom e.g plant, animal
3. motion in air, on land or in water	phylum subphylum e.g vertebrate
4. land: motor car, lorry, van etc.	class: mammal
5. car	order: primate
6. make: Porsche	family: *Hominidae*
7. Carrera	genus: *Homo*
8. species of Carrera e.g Cabriolet	species: *Homo sapiens*

Carl Linnaeus, the founder of modern taxonomy, thought a 'species' was a group whose members could interbreed. Darwin, as opposed to Linnaeus, preferred to 'plasticise' his definition of species to what he saw as its cause - gradual change. He wrote *"I look at the term species as one arbitrarily given, for the sake of convenience, to a set of individuals closely resembling each other, and that it does not essentially differ from the term variety."* However, Darwin not only regarded varieties as incipient species but also proceeded to *extrapolate* on the hypothetical principle of unlimited plasticity. He *did* consider this, a process called micro-evolution with which everyone agrees, a minor stage of macro-evolution. In other words, he guessed that variation was a progressive rather than a constrained, cyclical process. The question is plasticity (extent of variation).

Such extrapolation, variation-*without*-theme, formed the very basis of his theory. Through materialism's prism life is, naturally, progressive from a simple start. This, unlimited plasticity, is unquestionably the current scientific mind-set. Yet, we'll see, even the part of 'simple' start (called abiogenesis or chemical evolution) is fraught with intractable problems.[47]

Bottom-up, Unlimited Plasticity

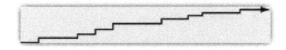

Top-down, Limited Plasticity

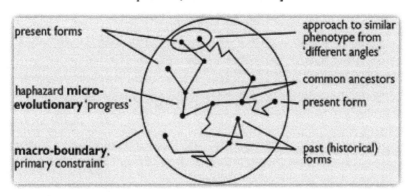

In practice, when genetics decisively demonstrates the limits past which mutation does not survive it also demonstrates conclusively, through myriad experiments (e.g. with *E. coli* bacteria, fruit flies, malarial parasites, flowers and mice), that the macro-evolutionary principle of unlimited plasticity (or unbridled extrapolation) is incorrect. Ask a dog-breeder if he has bred other than a dog. Ask him anyone has ever, by intelligent selection, pushed canine elasticity into the production of other than crippled or still-born, monstrously deformed dogs. No-one has made another type of organism from a wolf. **This demonstrates a limited plasticity.**

Thus, where Darwin staked his claim on continuity, *discontinuity* (the primary prediction of an archetypal model) is what we actually find. Nobody is more aware than a *palaeontologist* that Darwin's 'interminable varieties' are missing from life's petrified family album. And if, of course, the truth is that types are ring-fenced and discontinuous. Family trees are separate.

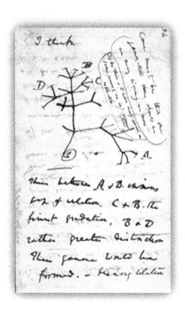

Darwin's inspirational doodle isn't right. The central assumption of his tree of life is that homologous molecular or morphological patterns will, by comparison in different organisms, illustrate degrees of phylogenetic relationship. Hasn't this idea been uprooted, sawn and cast as dust by an 'onslaught of negative evidence'?

Chemical evolution is demonstrably a non-event yet study attracts much taxpayers money (in the form of grants and university tenureships) around the world. Why? Because it is the seed, the root of evolution's tree. If the axe were laid here why not elsewhere? I have devoted 2 chapters of *SAS* to giving over 25 cumulative reasons why such alchemy is, except to the most wishful of speculators, impossible. **A couple, in a process that, as well as energy metabolism, had to be present correctly codified in the first cell. Transcription and translation through a ribosome of *DNA*. The process, multi-stage and multi-component and self-coded-for, must be present from the start of bio-informative operations.** In this case, why is the rational notion of an archetype so 'repellent' or 'plain stupid'? It is materialism and, especially, its subset of scientific atheism that need to damn it so.

The human genome could be stored on enough USBs to fill 2000 Titanics. Its 250,000 or so pages of close-written lines of letters would compose a book weighing 450 kilos. This code needs be accurate. At the other end of biological scale biologist Craig Venter defined a minimal genome by removing genes, one by one, from the already very small genome of a *Mycoplasma* bacterium. If the organism survived he discarded the gene until he was left with a minimal genome of 473 vital genes. But symbiotic *Carsonella rudii* has certainly lost the bare necessities of life since its present genome of 182 genes (~160000 'letters') is insufficient to replicate or transcribe and synthesize protein. A viable analogy for free-living *LUCA* might be ten times that amount, that is, 1.6 million letters (the size of a 600-page book) in the precise order of a program that will create specific molecules and shapes. Are natural selection and mutation up to the task of sufficiently 'improving' the survival chances of a highly speculative first cell?

First, historically, let's take Darwin's genetic editor called **natural selection**.[48] Edward Blyth (1810-73), a keen naturalist and 'Father of Indian Ornithology', published essays that first appeared in The Magazine of Natural History in 1835, 1836 and 1837. *These, which were read by Charles Darwin who afterwards corresponded with him, introduced the ideas of a struggle for existence, variation, natural selection and sexual selection.* These four are central Darwinian tenets and we might ask why, since in large part Darwin's work was based on Blyth's ideas, the former hardly acknowledged the latter's 'intellectual copyright'. And Alfred Wallace (1823-1913), co-founder of the theory of evolution and whose paper 'On the Tendency of Varieties to Depart Indefinitely from the Original Type' preceded (1858) and precipitated publication of Darwin's 'Origin of Species' (1859), noted that natural selection not so much selects special variations as exterminates the most unfavourable. Such extermination is, in fact, a stabiliser; 'fit' traits blossom, 'unfit' wither as an organism's own ecology (its niche) dictates.

So let's treat editorial selection for what it actually is and, at the start of our inspection of its 'mechanism', post three *caveats*.

The first, Blyth's, is that natural selection is solely a process of elimination. It involves the disappearance of 'unfit' organisms. **In short, deletion is not creation. Natural selection originates absolutely nothing at all.** It is, simply, a fateful finger hovering above a genetic delete button, no more than a name given to the lucky survival or unfortunate death of an organism or group of organisms in a particular environment.

Secondly, the effect of natural selection is, in the case of speciation, to have *reduced* the genetic potential of an original gene pool. Alleles are deleted or a pool split into separate populations. Information is not gained but lost. *For evolution, which needs not information ($\downarrow$) loss but ($\uparrow$) gain, this is entirely the wrong direction.*

Thirdly, it is false to conceive of natural selection as a kind of 'ratchet' that holds on to 'an extrapolation of improvements' at any level from nucleotide through protein to whole body shape. This is because without a prior plan to work towards there is no way for intelligence, let alone total lack of intelligence, to 'discriminate' a 'good' apart from 'bad' move as regards some novelty. *Indeed, it will breed out unwanted, nascent or non-functional characteristics.* **Only once an organ or a functional system (such as a beak or eye and associated factors) is complete and fitly working can natural selection act on trivial, accidental variation to that system**. **In a phrase, the 'mechanism' explains survival not arrival of the fittest.** It weeds the weaker but cannot create the fitter; it's as creative as a kitchen sieve.

In short, this mindless editor confers not change but stability by maintaining wild-type pedigree. Far from being a grand 'law of nature' natural selection is a trivial observation - an organism born defective does not reproduce. Its truism states the obvious. 'Weaker die, stronger live'. 'Who survives, survives'.

Mutation?[49] In 1865 Gregor Mendel presented his Theory of Discrete Units of Heredity. This was picked up by the Royal and Linnaean Societies in London and (since 'adamantine particles' are inflexible and anti-evolutionary) lost until, in 1901 de Vries discovered a way out - **mutation**. Henceforth random mutation became the innovator, the creator on which natural selection might, by killing, work. **Materialistic faith is vested in a core of unpredictability, a central lack of reason - mindless chance.** A scientific G.O.D, no less! A Generator of Diversity! By philosophical necessity current science takes its chance on chance. Materialism's Lady Luck, however weak, is elevated to almighty creativity, creator of life forms.

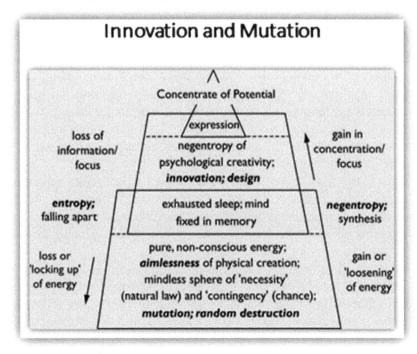

Innovation and Mutation

∧
Concentrate of Potential

	expression	
loss of information/ focus	negentropy of psychological creativity; *innovation; design*	gain in concentration/ focus
entropy; falling apart	exhausted sleep; mind fixed in memory	negentropy; synthesis
loss or 'locking up' of energy	pure, non-conscious energy; *aimlessness* of physical creation; mindless sphere of 'necessity' (natural law) and 'contingency' (chance); *mutation; random destruction*	gain or 'loosening' of energy

tam/raj	Sat
imbalance	Balance
system	Governance
variation-on-theme	Archetype
mutability	Immutability
informed	Informant
↓ tam	raj ↑
anti-life tendency	life tendency
incoherence/randomness	systematic coherence
accidental disintegration	purposive integration
cacophony	harmony
error	accuracy
babble/ 'noise'	code
meaninglessness	meaning
incorrect information	correct information
genetic/functional loss	genetic/functional stability
damage/disease	health
disorderly recombination	orderly recombination
mutation/misprint	facsimile

After all, mightn't mutations, if you found millions of 'beneficial' ones, transform one type of body to another in the way accretive evolution needs? *'Beneficial mutations' (BMs) are to genetics as 'missing links' to evolutionary palaeontology - critical.* **Not just The Primary Corollary but The Primary Axiom of Materialism and its whole panoply of philosophical, political and sociological speculation, not to mention the paradigm of modern science, hangs on their slender thread.** *The whole of secular academy depends, for its verbose existence, on this evanescent gleam of hope, a key to unlock all of life's diversity that's called a BM ('positive' or 'beneficial mutation'). Can BMs rise to such occasion? Everything depends on this.*

In fact, any *BM* would be invisible at the level of a whole organism and its small 'advance' overwhelmed by neutral or deleterious mutations long before the successive and exact chain of cooperative *BM*s needed for any novel biological system could 'evolve'.

It applies to 'junk', 'neutral' or any other kind of *DNA*. What should a first or following *BM* be? One can only be defined within a preconception. What is 'good' or what is 'bad' is only so when valued in anticipation of that end. **How can you even define a *BM* if you don't know where you're going?** Mindless evolution's natural selection doesn't know. And if there's no such thing as target how can 'progress' gradually mint a system with its integrated working parts? **With such lack of logic bang goes the irrational *PCM* and therefore bang goes *PAM* as well!**

Not knowing where you're going is a fundamental problem. The fact is that *DNA* represents, effectively, an organism's program. Now, in 2018, sequencing machines can tell you the order of nucleotide 'letters' in the text. This leaves us with an instance of Champollion's much more complex problem in the decipherment of Egyptian hieroglyphs - how to translate what the program is saying/ doing. By now we understand that certain sequences lead to the construction of certain proteins which form part of metabolic, intracellular and larger, extracellular systems. We do not understand the dynamic of this program (which is rather like trying to read the operation of a computer system by means of tracking hardware signals); nor do we understand the integrated operation of the multitudinous switching between protein and *RNA* subroutines. Progress is impressive, slow and fills the pages of prestigious scientific journals with its perceived complexity but only a start has been made.

Does *DNA* operate like a digital computer program?[50] There are close resemblances. And every systems engineer knows that a change to his construction (whether accidental or deliberate) either brings things to a grinding halt or causes ripple effects that, unless properly balanced and calibrated with cooperative parts elsewhere, degrade the functional intention of his design. Similarly with software. Random changes without reason or systematic realignments have, at best, no effect or, at worst, bring catastrophic failure. They never improve it!

In short, we understand that genetic information is at the root of biological form creation and maintenance but not how the program 'knows' how to build objects and events that make precisely the right parts at the right times in the right quantities and places. We do not know exactly how this automated factory program generates development (which involves anticipation) of any tissue, organ, system or whole body. When we do we will be able to say that we thoroughly understand this or that machine in all its aspects. But will this mean, as a materialist might claim, that we completely understand?

As Michael Polyani argued, machines are irreducible to physics and chemistry. They are irreducible because they involve immaterial purpose, the stepwise development of a plan of implementation, a directed cohesion of working parts and, of course, the thoroughly non-material anticipation of an operational outcome. For example, to completely analyse a bicycle you have found or dug up does not mean you have completely understood that form. You need to include its purpose, conception and technological development, that is, the mind in it. You could make stories about the evolution of its parts (including, perhaps, wheels homologous to other wheeled machinery) but simple physico-chemical analysis would never by itself obtain more than a fraction of the whole truth.

Yet, from the depth of ignorance, it was touted throughout the 20th century that we knew such bio-programs whose operation we still hardly understand evolved by chance mutations acted on by natural selection. Chance mutations '*must have*' caused the code with its adaptive, switching systems but, after this appeal, it is never explained exactly how (in what combinations or algorithmic steps) or why such mutation was, although at that stage neutral or harmful (i.e useless or worse), on the way towards some 'improvement'. **This is not science. It amounts to the multiple invention and repetition of just-so stories.**

In short, such materialistic explanation of the presence of life on earth, which concentrates on informative gibberish rather than excellence, is a large amalgam of stories that are basically irrational because machines (including bio-machines) are always the product of reason(s) and cannot appear by accident. All the scientifically inventive stories are engaged to subvert this truth.

Perhaps a deferential nod to theoretical probability in vast, vague time is an Oz-like answer, a veil that might irrefutably, unaccountably evade rationality, cancel out impossibility and seem to keep the faith alive.[51] However, at Cambridge Douglas Axe experimentally discovered that the number of changes needed to produce a novel protein-fold exceeds those whose degradation will deliver it to natural selection's lethal arms. He estimated that the probability of a 'mutational trial' generating a single, specific, 150-residue operational protein is 1 in 10^{77}. This, even allowing for several working possibilities, is the measure by which, in vast 'combinatorial space', 'gibberish' outweighs 'operational sense'. You might therefore, most reasonably assert (since you'd need other changes simultaneously to build more novel proteins, simple metabolic pathways or the many other complex items even very simple cells involve) that sheer Mount Improbable cannot be mastered - not, at least, in gradual, evolutionary ways. Oz is a ghost. **Darwinian theory is statistically dead.**

Yes, variation-in-action certainly exists. All agree upon the limited plasticity of micro-evolution. And on adaptation to fresh circumstance. Could not, however, an adaptive potential be written into genome and its epigenome so that it would not be mutation but a flexibility of program that produced coloration changes, different beak sizes and so on? Is speciation or micro-evolution really evidence for transformist macro-evolution?

In science you experiment. Can you test evolution in the past? In his book 'The Edge of Evolution' Michael Behe demonstrates that nature has empirically tested Darwin. Do you want numbers showing how, by gradual mutations, life might step-by-step evolve a tree? *HIV, E. coli* and malarial parasites satisfy the numbers game. Virus, bacterium and eukaryotic cell have reproduced, mutated and should have evolved through sufficient generations with sufficiently large populations to indicate whether neo-Darwinism's engine, random mutation, can bear the weight of a theory that would have it gradually innovate parts and body-plans by gradual but cumulative, useful steps - or, buckling, is crushed by numbers.[52]

In short, then, micro-evolution is a prejudicial, biased word because it implies the existence of an extended process for which no hard evidence exists **The reverse. Macro-evolution is a theoretical phantom.** In this view, **evolution-in-action is simply variation-on-predesignated-theme**; this variation is always constrained by working systems already in place; and it is either coherent, due to in-built genetic potential or incoherent by Darwinian mutation. *Variation proliferates; the special theory (STE) is right. But the general theory of grand macro-evolution (GTE) is not.*

Energetic matter knows no future. It cannot program, plan or anticipate. Yet, informant egg to informed adult, reproduction shows anticipation. Target. Purpose. Reproduction[53] defies entropy and death. It means the type survives.

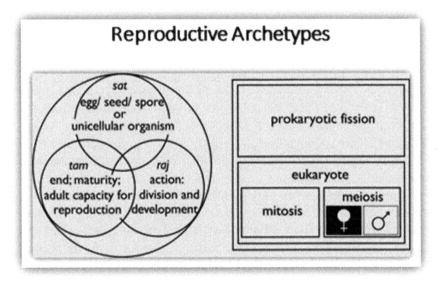

And due to lack of time we take, of types, meiotic sex. The biological point of sex is variation-on-theme. It is neither to add nor subtract but to shuffle an organism's cards into new permutations and thereby deal new hands in an old game. In this case two become a different one. Sex involves clear anticipation, targets and complex programming - not features normally attributed to mindless matter or to chance. How did it arise?

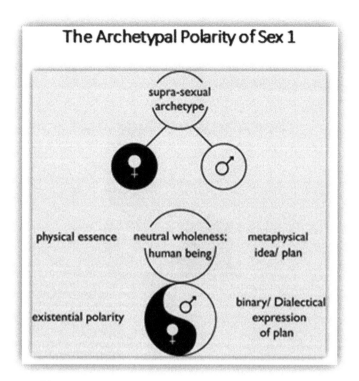

The Archetypal Polarity of Sex 1

supra-sexual archetype

physical essence   neutral wholeness; human being   metaphysical idea/ plan

existential polarity   binary/ Dialectical expression of plan

And sex?[54] If you wanted, theoretically, to design complementary sexes you might opt to specify each module using thousands of genes; *alternatively, and far smarter, you might conceive a main routine which included a 'gender switch'. Each type of life form that exhibits sex is conceived of as a neutral whole divided into polar male and female sexual halves*. **The sexual archetype (<u>not</u> a common ancestor) and its *DNA* coding are fundamentally hermaphrodite and incorporate both male and female potential**. Such hermaphroditic concept's switch would trip a male or female line; it would call gender-informed sub-routines. In principle, therefore, at a mere flick the balance would be tipped. A hierarchical cascade of emphasis would ensure dimorphic forms occur. **This skilful concept, this consummately programmed switch is precisely what we find in bio-practice.** From the same cells grow, in each gender's case, male and female parts. For example, sensibly perceived, male breastless and female breasted nipple are, as penis and clitoris, examples of the differential expression of the human archetype. Indeed, dimorphic *and* hermaphroditic (but never multimorphic) sexual algorithms are found in plants and animals. From hermaphroditic potential derive uni-morphic hermaphrodite, di-morphic male/ female and other forms of expression such as, by successively alternating sexual *and* asexual forms, even manage to polarize the reproductive archetype itself. You know for certain, therefore, the same program can deliver, for example, sperm (including the set of over 600 proteins that compose its eukaryotic flagellum) and egg (with her own specialities). Of course, genetic switching *aberrations* (e.g. intersex, abnormal hermaphrodites or trans-sexual tendencies) may occur as well but, normally, sex is implicit in a neutral archetype's potentiality.

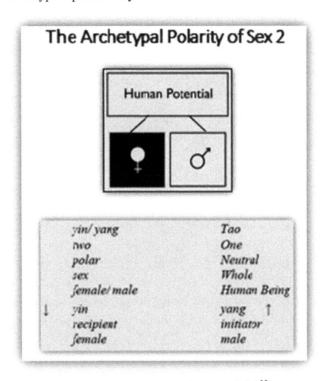

And as for eggs, did an egg precede its adult?[55] You can't see a memory but have you ever seen a physical idea? Ideas are potential; from potential outcomes are evolved; an egg, packed with symbolic information, is as near to 'natural idea' as you will get.

Did an adult precede its egg? Did fruit precede its branch? We noted that reproduction looks to the future; and thus sex *anticipates*. It is a conceptual process engaging machinery of irreducible complexity to achieve a target - generation after generation. Survival of the kind.

Egg and Adult Together. What has been noted bears repetition. **Wherever an apparent 'chicken-and-egg' situation crops up the puzzle is resolved by the introduction of purpose, design, information and mind rolled into one - teleology.**

As far as chicken and egg are concerned, therefore, the simple answer of an information technologist is that neither came first. One did not precede the other. You make them together with the same object in mind - in this case a self-reproducible biological information system or, in other words, a living being. **You are one half of a metaphysical idea called 'human being'.**

Meiosis, which underwrites the sexual reproduction of some single-celled and nearly all multicellular organisms, bears the hallmarks of a choreographed routine designed to extract maximum variation-on-theme at minimum cost in labour and materials.

Meiotically generated gametes then, by means of and within a highly specified and complex context, fuse; and their offspring, a single-celled zygote, unfolds according to complex but codified, specific algorithms, a hierarchical developmental archetype, which eventually realise a goal, re-creation of the next adult generation.

The Order of Development 1

surrounding field-of-action	*Axis*
physical expression	*Plan*
material outcome	*Immaterial Concept*
↓ *detail*	*outline* ↑
lower-level expressions	*top-level guidance*
informed system	*informant system*
micro-individuality	*macro-generality*
minor instability	*major stability*
variation	*invariance*
target parts	*symbolic representatives*
codified product	*generative code*

As for that supreme natural example of potential expressed, targeted plan and codified *top-down* program, development[56], how would *you* have done it? If *you* needed to achieve the imperative of survival down the generations, how would you have copies made? Especially if you wanted to include some sober flexibility, that is, continually varied reproductions on a standard theme? *We can ask how an engineer might design a self-reproducing machine and ask if nature has preceded him in his intelligent logic. How might the least demand be made on the tissue or strength of a parent while at the same time encapsulating its potential?* <u>*A brilliant, optimally economical idea would be to reduce the parent to a single cell and then, from the symbolism of this cell, build up a new adult.*</u> This is exactly what happens in nature. Each individual adult is 're-potentiated'. Its body is reduced almost entirely to a symbol, a directory, a coded book of what might be. Its spring is re-compressed into the top-level potential of an egg or sperm. A double dance is at the heart of life's sexual gamble.

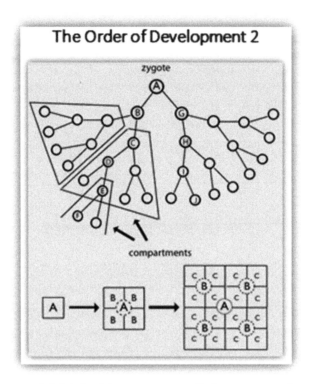

The Order of Development 2

Developmental biologists uncover switch-arrays and associated, complex, hierarchical cascades of regulation. They reveal top-level 'master genes' or 'tool-kit factors' that determine which parts of a body will develop out of others. These blocks of levers clustered as a signal box that rules developmental lines remind you of a railway running from a start to terminus. They remind you also of computer software. **For example, a cluster of control genes would represent a top-level or main routine from which genetic sub-routines were switched; *Hox*, *Pax* (for nervous system including eye genes), *Otx* gene family (for kidney, guts, brain, gonads) and other 'informative centralities' would, with the brilliant Boolean logic of a computer program, mastermind the developmental output of bodies ranging from a jellyfish or fly to you - with a somewhat different kind of logic to govern plants and fungi!**

Regulatory logic-clusters are found, it will predictably transpire, in all multicellular animals. For example, in four distinct blocks in vertebrates (called *Hox* gene clusters) each gene is responsible for triggering a cascade of sub-routines that will supply the right materials in the right place at the right time to construct a given segment of the animal in question. Obviously, therefore, a *Hox* gene for a worm will define (or, in computing terms, call) different or deeply differentiated organs (sub-routines) to ones for a fly, a horse or a human. In other words, the same genes can be responsible for initiating the

116

development, in terms of systems or organs, of different outcomes. In one instance a gene may specify for a tail, a coccyx or the rear part of a fly, frog or grasshopper; and the gene that triggers development of your eye would, if transferred to a fly that lacked it, cause its blindness to be eyed - with a fly eye not a human one, of course!

How does chance evolve a path that targets goal? How, entirely ignorant of end-game, did as-yet-useless intermediates survive? Missing links die of incompetence thus how did any metabolic pathway prophesy its own construction or feedback control; how (as, analogously, in the logical, consequential proof of a mathematical theorem) did thousands of correct steps on the path to reproductive adulthood arise by accident? Evolutionary 'must-have-somehow', 'just-so' stunts are pulled continually except, with metamorphosis,[57] the bluff is called spectacularly. Butterflies have always thrown Darwinism, in an arm-lock, in a flap and on its back.

Back-to-front. The transformation from egg to beetle, bee, fly or, more strikingly, a butterfly illustrates a pattern of development that defies a gradual, practice-to-principle evolutionary explanation. Entirely different-looking phases, each perfectly formed for function, serially erupt. First *egg*; then, for a 'childish' *caterpillar* to become an adult moth or butterfly, it eats and moults. It keeps moulting exoskeletons (which are flexible like cat-suits and yet give it shape) for larger ones that form folded underneath the smaller outer sheath. At the last and largest size skin is shed by delicate manoeuvres revealing a cocoon, a *chrysalis* in which the future hangs. Then caterpillar body parts dissolve and build again into a butterfly called the *imago* that, emerging after several days, inflates its lovely wings by pumping blood into their veins.

Could humbly waiting as a caterpillar eventually evolve a program (saved as what are called 'imaginal discs') for pupal development? In what nascent cocoon could enzymes 'know' how far they should dissolve a caterpillar's parts before re-building to transfiguration called a healthy butterfly? How long did some ancestral pulp hang round inside a chrysalis (that came from who knows where) until a suite of chance mutations magically (how else?) re-programmed 'mush' into the concept of winged flight? How, in fact, did pupal death rise straightway (not even in a generation) to a form by which type-butterfly appears? One must ask also how, without the benefit of plan to reach anticipated adult form, serial immature phases took hundreds of millions of years to accrete. *It is noted that before any organism could 'create' any new stage of development it would have first to reach its present stage's adult limit then add that new stage.* This is because it has to reproduce to create the fresh, 'advanced' offspring. A reproducing caterpillar? Such back-to-front order is patently absurd.

From principle to practice, conceptual information to biology - imaginal discs, homeobox genes and other features simply demonstrate the *top-down* venture that, at climax, claps on stage a butterfly.

Butterflies are symbols of nemesis. At the clap of silent, fragile wings Darwinism logically dies; a giant is slain and at the same time flutter flags of life's innate, original intelligence. She lifts oceans then exactly drops them down again. The effect is not so large that continents are flooded and eroded each high tide and not so small that surface waters can stagnate. Marine environments are flushed and with diurnal freshness the communities of ocean bloom and thrive. Lunar periods influence other biological events and our consort (in company with vaster Jupiter) also shields mother earth from a barrage of comets and asteroids.

Signs of obvious anticipation always flatten evolution. They squeeze the theory's time to death and thus compress it to impossibility. How can forethought be by chance? When is a plan not a plan? Is concept the same as lack of it? Biological evolution entirely *lacks* concept or target yet its theory equally lacks any serious explanation how, through many specified and integrated stages serial *and* parallel, the evolution of development '*must have*' occurred. Is this 'rationalism's' finest hour? **The fact is that all instances of metamorphosis (of which development in general is one) are another Darwinian black box.**

Sex, metabolism, metamorphosis and morphogenesis are four <u>*anticipatory*</u> *and therefore conceptual processes.* **The fact is that the evolution of development is as much a black box as the evolution of any of them. Their bio-logic hammers nails squarely into at least four corners of Darwinism's and thereby materialism's coffin.**

In summary, the choice of chance as G.O.D (generator of diversity) is perverse. There is no doubt that biological systems are informatively codified. They are complex, hierarchically organised, targeted and, like hierarchical, targeted computer systems, as intelligently conceived as they are self-consistently arranged.

Lecture 6:   Community

The system of Natural Dialectic is, as much as being an abstract reflection of the way things are, an application program. This program generates a template for both involuntary, instinctive and voluntary, chosen behaviours. It organises the pattern not only of 'hard science' and biology but politics, law and religion; it guides the aspirations of education; it is hard-wired into the humanities and, as such, is expressed in the very fabric of individual and social life. Its consequences, especially moral and psychological consequences, involve everyone. How?

We're going to work towards a Unified Theory of Community.[58] **From a bottom-up perspective humans are animals and mind evolved as a strange function of brains. Top-down,** *however, the first compass of community is universal and absolute.* It reflects a structure of creation whereby all things, psychological and physical, descend from an Absolute Source. In this sense alone is everything connected or, as some aver, 'is one'. **Such Absolute Community of Essence and Existence is illustrated in this diagram.**[59]

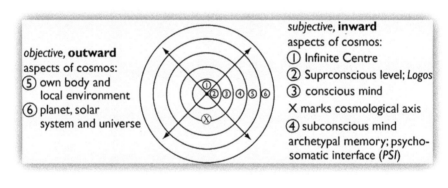

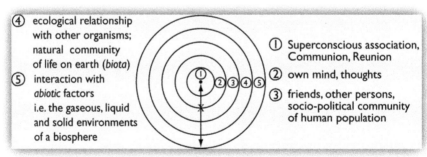

However, the Relative Community of Existence from a Standpoint of X (circles inward and outward from X) recalls the cosmological

bearings by which we first oriented ourselves in Lecture 1.[60] From this point X, however, social circles radiate in *both* directions.

Internally there exist the symbolic yet most real relationships of mind. They involve other persons, bio-forms, events, objects and the paraphernalia of a life spent learning. This is the experience of your mind-world.

Externally you descend from X through physical interaction with your environment. The rings spread first through your nervous system (which is the closest physical associate of mind) to the rest of your skin and bones and their neighbourhood. This is your space-time location. It includes other biological bodies to which you physically and, more or less easily according to type, psychologically relate. If you restrict these to the human type then you involve the dynamic of family, friends, neighbours, interest groups, nations or the international scene. You take a social part, which may involve institutionalised politics, law and religion, in your community.

More broadly, you can think of life-forms (biotic factors) in terms of individuals, populations or communities of different kinds of organism. This positive, inclusive perspective is called ecological. **Thus ecology is about your 'wider body'.**[61] Such body includes a community of organisms (*biota*) that inhabit the surface region of our planet. This region, called an ecosphere, comprises living and non-living parts. The latter descend, as the rings convey, to include non-living elements (*abiota*) such as sources of energy, climate, ocean, water, mineral cycles and the soils of earth. The compound amounts to a stage on which, in different scenes, various players act. An ecological play is dynamic. In such a network the health and behaviour of each part affects its whole.

Firstly let's survey **non-living factors**.

Not only our galactic but also our planetary zone is habitable. Life needs an endless supply of liquid water and thus strict, natural thermoregulation. Our thermal generator is a 'dwarf main sequence star'. Our lives are hung upon this lucky star. Precise strength and character of the four basic physical forces keep it burning radiantly. Not only earth's distance from the sun but also its unusually circular orbit slung on a fixed radius and a specific kind of rotation round its own axis is each exactly felicitous. They're appropriate by distance because temperature on the globe's surface is, except at the poles, life-friendly; by circular orbit because this temperature is stabilised; and by speed of spin that avoids life-destructive boiling, refrigeration and violent, protracted wind-storms. Critically, for billions of years at least, maintenance of constant temperature at precise degree has produced that earthly bio-zone - a liquid water-bath in whose fluids carbon-based molecules vital for life are stable so that bio-forms can thrive.

If life's solar lord clocks fine statistics what about our closer lady of the night? Four hundred times smaller but four hundred times closer, at the moment of eclipse she exactly covers him. With us she dances near enough to lock in motion so she never shows a dark side, just a single, silver, sunlit face. Our planet's tilt, which wobbles through the seasons and affects sea levels, has been finely damped and stabilised by moon's gravitational effect.

Pure energy (sunlight with harmful frequencies deflected or deleted) falls on gas. Earth's primary dynamic, sky, consists of a concentric suite of atmospheric shells. Each, like a membrane, offers its particular aegis to the life within. The living planet floats, egg-like in a white of air, within the warm, deep womb of solar influence. You might construe the stratosphere and ionosphere as buffers, membranes, even subtle skin.

Nested lower down the suite of air-light shells you find fluids of fertility - mists, rain, rivers and the oceans. These, lower atmosphere and ground water, assume the blood-like role of convectors, radiators and conductors. The amount of water on the blue planet has remained stable. So when rivers pour megatons of salts into the oceans how do these avoid acceleration into dead seas?

The Mother's bones, nails and hair are soils and solid, crustal rocks that, washed by storm and stream, yield minerals. These minerals life's producers, plants, absorb. Volcanoes throw up irregular formations like mountain chains whose various habitats permit an abundance of ecological niches. If life's sac is sky, the earth's skin is a crust of islands (or tectonic plates) that float on seas of magma.

Indeed, animate is coupled with inanimate. A system is a network of ideas or objects linked by common purpose. *By this definition life on earth is called an ecosystem.* An ecosystem includes living and non-living factors combined into a single self-regulating system. For James Lovelock's Gaia theory earth is a 'super-organism' made of all lives tightly coupled with air, oceans and surface rocks. Such a 'super-organism' maintains dynamic equilibrium. It comprises a totality that *seeks*, in a cybernetic manner, optimal conditions for life. Its variables include temperature, pH, salinity, electrical potentials etc. Biological controls include metabolic, hormonal and nervous systems of adjustment. In this case the *reason* for such coherent operation is not, in the last analysis, physical but metaphysical. **'Cybernetic homeostasis', like 'program', is a conceptual phenomenon. Its presence in any machine indicates an underlying purpose.** *Numerous biological sub-systems are coordinated under the overall purpose of the continuation of life, that is, of survival.*

For their biological part individuals, populations or communities of different kinds of organism are involved. Life's geo-physiological health, the poise on which all ecosystems and their multicellular inhabitants depend, is in great part the gift of bacteria. Whatever their mode of origin, microbes of the kinds that exist today always existed. In the sun and fresh, oxygenated air of the 'over-world' flourish aerobic types; in the dark, anoxic soils, sediments and mud of the 'under-world' slave multitudes of anaerobes. Tough and reliable, they toil relentlessly. They 'plough' the earth and continuously 'farm' organic substrates on which other organisms thrive. Bacteria might even, as the foundation of life's ecological pyramid, be construed as its primary, substantial, most important form of life; yet, working at the interface with inorganic matter, most 'robotic' too.

Provision, consumption, recycled waste. *You need all three.* Input, process, output. *Homeostasis needs all three.* Ecology is irreducibly, biochemically homeostatic and cyclical. Together every community and every level of life in each community cycles around each co-factor. Indeed, each organism plays one or more of the roles. You need three-in-one to peg the balance happily.

Yet now one actor has claimed centre stage. Happy balance is upset. Monocultures easily destabilise; they weaken linkages that make life's safety net so strong. *Indeed, at this point an acute paradox rears its head.* **Man - whose intelligence, forethought and powerful creativity should serve in the role of steward and conservator of life on earth - seems to have lost his better mind.**

Let's turn, before considering deliberate negativity, to nature's shows of seeming negativity.[62] An ill wind, evil cold, cruel sea and other natural challenges (not least, inescapably, the body's own calamities) may threaten life with suffering. They may spell fearful pain and death. Such 'evil', as we understand, does not involve intent or animosity. It is, as matter is, oblivious: not immoral but amoral. Thus 'ill wind' does not blow with ill intent. It blows according to the fashion of inanimate design. In short, nature's so-called 'negativity' is really its insensible neutrality.

We pray for relief from personal catastrophes. We stamp our foot because we don't get all we want the moment we demand it; or make demands that counter natural law; or curse because we can't conjure suspension of its operation? Shall we rationally blame an engineer whose locomotive knocks us down? Isn't this, which many prayers adopt, a childish attitude? If every one of nature's children kept on wanting different things, then how confounded would that nature be! There would be no end to muddle. Confusion, chaos and confliction of the elements! Yet humans, when we suffer, pray such 'imperfection' be perfected while cursing the imperfect part. We cry, unrealistically, for paradise on earth.

Voluntary negativity is quite another thing. Evil deals, intentionally, in lies and pain. Its condemnation isolates, its burden weighs you down. Negativity inflicted on another is the state of hell.

Yet God and suffering can, in a polar, vectored cosmos, logically co-exist. Why not? How could the world be always right from everybody's point of view? But if there is a God of love why should He create an entity of malice, cunning and perversity? Is, therefore, God father of the devil and, thereby, the root of evil? Who loosed the wolf of persecution on a world of suffering innocents?

all below	*Transcendence*
lesser sorts of being	*Supreme Being*
vectored deeds	*Potential/ Poise*
balancing acts/ reaction	*Balance/ Pivot*
range	*Super-State*
range of shadows	*Essential Light*
moral spectrum	*Good*
↓ *negative act*	*positive act* ↑
division/ demonization	*unification*
from Truth	*towards Truth*
from Peace	*towards Peace*
body/ self-centred	*soul-centred*
passion	*detachment*
contraction	*expansion*
malevolence	*benevolence*
hate/ abuse	*love/ care*
crooked/ perverted	*straight/ open*
criminal/ demonic	*saintly*
a curse	*a blessing*
immorality	*morality*
descent/ darker	*ascent/ lighter*
negative wish	*positive wish*
tendency to create chaos/ cause pain	*tendency to order/ pain relief*
decline/ fall	*lift/ helping hand*
depriving	*sharing*
malefactor/ enemy	*benefactor*
evil/ sin	*goodness/ virtue*
darkness	*light*

Resistance, opposition and exhaustion - the world exists through shadows set against its light. Beneath transcendence truth is broken by polarity into opposing vectors; plus and minus are unbreakable a pair. But as free agents and not unconscious robots humans *choose* which way to act - and sometimes voluntarily create another's pain. Holistically-speaking, don't blame divine but human choice for evil. Thus, if evil's sourced in men's own heads, moral struggle isn't cosmic but just local.

Who denies that in a fallen world the bestial side of life is struggle for survival? 'Survivorship' and 'reproductive fitness' win a day that's governed by genetic products called your limbic system and its master glands. To rape, cheat, pillage and exploit is thus implicit in your naturally selected genes.

What, therefore, might a well-evolved and cunning despot's genes inflict upon 'the enemies of scientific reason'? Deceive, purge, pulverise? Eradicate all threats to his genetic egotism's article of faith - any-cost survival? Ruthless, forceful and, the best of all, efficient tyrants soon create such 'excellence' as Dante's hell.

Racism. Humans are genetically 99.9% or more identical. If they descended from an original pair they would obviously all be, in this sense, blood relatives. Brothers and sisters. Darwin's title is, however 'The Origin of Species by Means of Natural Selection or The Preservation of Favoured Races in the Struggle for Life'. Clearly, in this context 'race' approximates to 'species' but, where 'race' is nuanced with an innate sense of superiority-and-inferiority, Darwin was, with the great majority of his contemporaries, of a supremacist mind-set.

Fascism. 'Higher race subjects to itself a lower… a right which we see in nature.' Hitler argues[63] that Darwinism is the only basis for a successful Germany; and in December 1941 he revealed to his Secretary, Martin Bormann, that his life's final task would be to solve the religious problem - the organised lie must be smashed. Franco and Mussolini also sprang from Nietzsche's *Übermensch*, apostolic Haeckel's febrile evolutionism and, therefore, once more Darwin's and his cousin's pseudoscientific corm. Indeed, world records for mass-murder germinated like black flowers of modern twentieth century evil from that bulb.

Communism. Stalin followed Lenin who followed Marx - for whom a signed copy of Darwin's book contained 'the basis in natural history for our views'. Mao Zedong, Pol Pot, Kim Il Sung and others descended from Marx by way of Stalin. Dialectical materialism is central to the thrust of soviet ideology. Its oppositional stance treats opposition to its chosen 'positive' as an excluded 'negative' with which it is at war and wishes to eliminate. Taking scientific atheism as 'positive' this antagonistic form of logic thereby demonizes metaphysic as prime enemy.

War. Some humans love to hunt and kill. War's the climax of such 'sport'. The devil's game. It inflames in-built passions to an internationally murderous degree until victory bestrides the spoilt and spoils. Evolution has, for the last 150 years, been pinned to martial and imperial masts. It may not directly instigate devilry but its theory, survival of the fittest, cannot logically condemn a fight.

From isolation let us turn towards a Unified Theory of Community[64] regarding social part.

How best, philosophers enquire, live life? What solvent best dissolves the harshness of our problems on hard earth, what solution

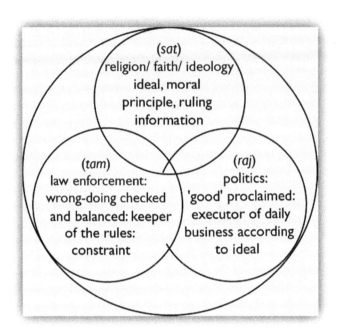

flows towards universal happiness? Does cosmic law (polarity) provide an answer even if it's not what *ego* might desire?

Down-to-earth solutions to the problems of disorder, ignorance and suffering involve two species of utopia - one objective, outward social and the other inward, personal and subjective. *We'll deal with the outward, large-scale solution first and then, as was the plan, work inwards towards a microcosmic, personal Nuclear Solution.*

Firstly, therefore, in considering a **Unified Theory of Religion, Politics and Law,** we speak about '*external association*'. Such association comprises relationships between an individual and his circumstance. It includes both positive and negative interaction with non-living objects, any living organism and, closest, other humans. At the heart of human community are religious, political, legal, educational, economic and personal communications with each other. The strongest, most important relationships are often with those in close physical proximity. This includes one's own body (habitually thought of as oneself), family, neighbours, colleagues and workmates. Of this group the closest are those 'on the same wavelength', that is, perhaps family but certainly friends. What*, therefore, should the instruments of social cohesion - law, politics and religion - most attentively and sensibly nurture but family and, within the family, its individuals. Is this your society's case?*

Religion, politics and law - this homeostatic triplex guides the individual and, by extension, his society. Is not the foremost regulator of the three, from which ideals derive, world-view? Secular or, if immaterial precedes and governs the material state of man, non-secular religion sets ideals. Ideals, as far as realised, compose desired utopia. In this case from Natural Dialectic's point of view best principle derives from Top Ideal. *Society's best formal answer is one that addresses evil issues from the State of Immateriality; our material solution is resolved, most definitely without recourse to science and without a test-tube anywhere in sight, by institutions charged with primary exercises in morality. Morality steps centre stage.* It is not scientific but it is for sure the bigger player. The purpose of the agencies of social order - religion, education, law and politics - is to combat imperfections, minimise all criminal behaviour and promote relationships. *The whole point of morality is maintenance of individual and thence social balance and, wherever necessary, restoration of dynamic peacefulness.*

body/ mind	*Soul*
lesser selves	*Self*
personae	*Psyche*
concentric rings	*Nucleus*
relativity	*Truth*
relatively sane	*Sane*
↓ *external*	*internal* ↑
descent	*ascent*
exit	*return*
lower self	*higher self*
from Creator	*to Creator*
from Centre	*towards Centre*
from Goodness	*towards Goodness*
criminal/ evil	*respectful/ good*
insane	*sane*

Religion means 'a system of belief' and, whether vague or clearly framed, is therefore unavoidable. Whether atheistic or theistic, oriental, occidental or plain secularly 'liberal' it frames relationship with cosmos.

For materialism there is only one way. There is no direction but its own wherein all life is evanescent and thought dances, like will o' the wisp, inexplicably upon the waters of oblivion. If this is hopeless and a counsel of despair grow up, the creed exclaims, grow strong because that's just the way this harsh, unfeeling, hopeless cosmos is.

If, on the other hand, you treat with Nuclear First Cause; you treat with Metaphysical Ideal, Transcendence and Top Teleology. **Mind's eye looks down upon a world of soul-less things but, swivelled**

127

upward towards the Apex, seeks illumination's jewel. You step up; and, in communion with Essential Soul, are transformed into supreme and living paradox. This we call nuclear religion.

However, mankind's peripheral religions, the world's formal faiths, are social encrustations born of ritual surrounding nuclear experience. Saintly teaching is expressed in differing, local ways. Holy scriptures, creeds, imaginations, theologians, uniforms and prayer-houses represent an inexperienced congregation. At best this expression points upward towards a knowledge of the Peak Ideal. At worst, however, creeds breed isolation. An oft-bellicose delusion of any particular organised religion is that it's 'the only way' or 'sole repository of truth'.

existence	*Essence*
spectrum	*Source*
hierarchy of riders	*Ideal*
lower principles	*First Principle*
consequent development	*Nucleus*
belief	*Knowledge*
religious faith	*Super-religious Fact*
world-tree	*Archetypal Seed*
↓ *darkness*	*light* ↑
peripheral formalities	*approach to nuclear core*
bondage	*freedom*
lack of intelligence	*intelligence*
rigid adherence/ dogma	*thoughtfulness/ debate*
hypocrisy	*sincerity*
rule by fear	*rule by wisdom*
arrogance/ intolerance	*toleration*
wrong	*right*

For a peripheral religious believer explanation of his faith is not required; and for what dogma forbids none will suffice. For atheistic 'infidels' the religious delusion is that metaphysic has existence. For theistic believers various delusions stem from encrustation of ideals, misunderstandings, misinterpretations, fanatical conformities and their hypocrisies. 'My gang, your gang' mentality directly leads to conflict and, destructively, to war. Illumination is reversed. Ignorant perversion of the upturned approach to nuclear core are twisted round and downward into politics of fear and clerical control - the very opposite of Truth.

From ideal theory turn to the lower truths of egotistical pragmatics. Below the third-eye's balcony observe life's market-place. Survey the action, business and economy of body - seething crowds, cross-currents of humanity; turbulence of social ocean, individuals jostling with each other in a way that keeps (or fails to keep) the peace. What is a crowd but a community, communities a town and towns writ large across a wedge of continent? Individuals in society, men in a collective state, states dealing with each other in a way that keeps (or fails to keep) the peace - this is the cauldron, **human politics**, into which birth throws us where we boil. As with biological so with legal and political dynamics. **Resilient balance minimizes stress so that equilibration is, for body politic, the basis of its politics**. The purpose of a government is steering towards its stated goals. Are these not, in a human nutshell, education and, in providing for a healthy body, keeping healthy peace? *Its purpose is to protect its subjects' sense of equilibrium by maintenance of law and order in its realm.* Employing metaphysical ideal how best, philosophers enquire, to legislate?

Are conscience, self-control and an internal law to be preferred? Or don't they count and you invite a visit from the officers of law? Rules leash animals with reason; they frame your practices with principle. You are constrained within a mental box of regulation. If you flout the ideals that it represents then an **external leash** of conscience must restrain. Those officers detain and teach you with the pain of punishment. They might even lock you in a hard box called a cell.

The second of two species of utopia is inward, personal and subjective.[65] *We now work inwards from Point X towards the Real Deal, Nature's Inward Positivity.* No crime! No prisons, courts or punishments! Do such societies, outside monastic, still exist in which self-government eliminates the reckless, disrespectful element? Where, really, does self-government exist but mind? So crimelessness has great potential. Nature's Inward Positivity exists in every human but its exercise needs choice.

Delve further in. Take nuclear religion to your own (and everyone's) extreme. *Internal, individual association involves just one relationship.* How well do you live within yourself? How to realise The Real Deal - Your Self? Further in. Take nuclear religion to your own (and everyone's) extreme.

Return to Source. Nuclear Religion[66] is about (↑) return from the periphery of creation to its Natural Centre. Isn't the Aspiration of a human life, naturally inlaid but much neglected and distracted, to distil the mind's pollution to a pure distillate and thus, as every mystic always told you, reunite your life with Life? Essential Psychological Unification: Arrival Home.

No doubt science cannot, though a scientist being human can, understand the theory and practice of a science of the soul. But can you now grasp that, at the Heart of Individual Association, is Reunion - Reunion of the self with Quintessential Self?

Inward shifting; trans-religious movement, a possibility that's always been incorporated in the human frame, is mankind's future evolution. *Now, as much as in the past, the affairs of men have need of such a healthy frame of reference as Natural Dialectic's. Now, as ever, most prefer equivocation.* The remedy is hard to swallow; its regime is hard to follow. Its logic neither replaces nor competes with any 'orthodox' expressions of Absolute Morality, rather underlines their Centralising Mode.

existence	*Essence*
relative truth	*Truth*
duality	*Unity*
↓ *division*	*unification* ↑
material focus	*immaterial focus*
towards illusion	*towards Truth*
no thanks/ ingratitude	*thanks/ gratefulness*

In conclusion, I ask whether *information* may not be the hidden, immaterial factor that causes currently materialistic paradigms to shift. Whatever the case, I hope that you have enjoyed the exploration of a philosophical architecture whose *motif* is the binary pattern of Natural Dialectic. We have, considering both inanimate and animate, circled round the universe. *What is, from the evidence, the final line and highest of conclusions? What, at the most natural core of cosmos, is the nature of Truth? The choice, the faith is at the end between material chance and an Informative Creator.*

Glossary

A

adaptive potential: involves possible changes to super-coded switches and recombinant refinements intrinsic in the genomic program of any particular biological type (*SAS* Chapter 23).

allele: a sister gene; you have two copies of life's book, one from mother and the other from father, so that each gene from father has a correlate 'allele' from mother and *vice versa*.

anti-entropy: *see* **negentropy.**

archetype: basic plan, informative element; conceptual template; pattern in principle; instrument of fundamental 'note' or primordial shape; causative information in nature; 'law of form'; nature's script; Natural Dialectic's 'holographic' edge, omnipresent but invisible because it's metaphysical; the place where metaphysic and its physic meet; morphological attractor or field of influence in universal mind; prototype-in-mind (maybe related to Platonic ideas, Aristotelian entelechies and/ or Jungian archetypes); potential matter seen as hard a metaphysical reality as, say, particles are physical realities; program(s) naturally stored in cosmic memory - simple in terms of inanimate physical 'law' (of particles and forces), complex in terms of animate structure/ function/ behaviour; information stored in a typical mnemone; in biology, metaphysical correlate of biological type/ super-species that is physically expressed in code as *potential* form; abstract or metaphysical precursor; as thought is father to the deed or plan is prior to ordered action, so archetypes precede physical phenomena; pre-physical initial condition of matter; see also SAS Chapters 16, 17 and 19; also PGND

ATP: Adenosine TriPhosphate, life's standard bearer of chemical/ heat energy; a cell's agent of energy transmission; a biological 'match' or 'battery'; an active cell may discharge many thousand units of *ATP* per second to drive its metabolic machinery; these are recharged by respiration; *ATP* also plays a critical informative role in the transmission of nervous and possibly other signals.

B

base: significant component of a *DNA* nucleotide: a letter in 'the book of life': there are 4 bases in the genetic alphabet - A (adenine), G (guanine), C (cytosine) and T (thymine): in the case of *RNA* base T is replaced by U (uracil).

base pair: the conservative accuracy of genetic inheritance and the elegant construction of *DNA* are both dependent on a base-pairing rule viz. G pairs only with C and A with T (or U).

big bang: *see* **transcendent projection**

biomimetics: also known as biomimcry; fast-expanding field of scientific study and imitation of the biological production of codified substances and processes; research in order to better inform the processes of design (engineering) and technology (production) for human purposes.

black box: process or system whose workings are unknown.

C

chaos: a confusing notion with three main but disparate implications - emptiness, disorder and randomness; Greek word meaning chasm, emptiness or space; structureless 'profundity' that pre-existed cosmos; *prima materia*, *prakriti* or primordial energy structured by regulation of divinity, archetype or natural law; anti-principle of cosmos i.e. disorder; any case of actually or apparently random distribution or unpredictable behaviour; also apparently random but deterministic behaviour of systems (e.g. weather, electrical circuits or fluid dynamics) sensitive to initial conditions.

chloroplast: organelle in plant cells containing photosynthetic apparatus.

chromosome: a 'book' in the 'encyclopaedia' of life; the human genome contains 46 chromosomes.

cladistics: method of classification using diagrams called cladograms; organisms are collected into groups on the basis of shared (homologous) features; homologies are tallied and numerical rather than speculative, evolutionary/ phylogenetic links drawn up between organisms; cladism is thus a powerful, neutral, objectively detached tool of analysis and for this reason the technique enjoys growing popularity among the world's taxonomists.

code: the systematic arrangement of symbols to communicate a meaning; code always involves agreed elements of morphology (the form its symbols take), syntax (rules of arrangement) and semantics (meaning/ significance); without exception such prior agreement between sender (creator/ transmitter) and recipient involves intelligence.

Communion: the Christian term for mystic union (see also transcendence, *samadhi*, *nirvana*, holy grail); realisation of Universal Truth; Salvation; Absolution; core subjective experience; attainment of Supreme Being; Core Psychological Experience.

conscio-material dipole: illustrates the basic components of polar existence; informative and energetic components may be graphically modelled (as *fig.* 2.5) on vertical y- and horizontal x-axes respectively; the origin of such graph (zero information and zero energy) represents the cosmic sink - an abyss of space); sources of the couple extend from infinity towards this sink; Archetypal Potential scales from

Psycho-Logical, Informative First Cause through grades of mind to non-conscious zero (matter); and potential matter (first cause physical) drops from original concentration of ultra-heat to, again, zero; thus cosmos is viewed as the gradual embodiment of Uncreated Source; it represents a scale of possibilities expressed as typical yet individual forms; by this token embodied soul is subject to individual incorporations (psychological and physical); these relatively dynamic forms constitute its psychological and, in the biological case, bodily circumstance; Natural Dialectic simply models such a hierarchical description of polar creation by the use of spectrum, concentric rings and, step-wise, ziggurats (see also cosmos and creation below).

convergence: the tendency of unrelated organisms to evolve similar characteristics; in the case of *divergence* adaptation/ speciation from an original feature occurs (e.g. beaks of finches); *convergence*, involving the unrelated, mosaic occurrence of similar features (such as the camera eye, viviparity and thousands of other instances), runs counter to Darwinian expectation; it means that such codified features must have evolved independently many times over; evolutionary explanations of this profound yet ubiquitous puzzle may thus involve speculations such as appeal to non-random 'deep bio-structure', 'principles of evolution', 'morphological laws' or 'inevitablility' granted by imaginary natural laws of codification/ innovation; for a design theorist the bio-codification and engineering of 'convergent' forms derives from either an original use of modular programming or, in the case of so-called micro-evolutionary variation, from in-built adaptive potential flexibly but appropriately activated by genetic switches and epigenetic markers.

cosmic fundamentals: cosmic psychological and physical qualities; three basic states or tendencies; universal ingredients whose mixture is variously expressed in every object and event.

cosmological axis: human pivot; the point at which subjective and objective perception meet; eye-centre; third eye; thought centre; *ajna chakra.*

cosmos: physical universe, universal body; denotes orderly as opposed to chaotic process; involuntary pattern of nature; also equated, including metaphysical mind, with existence as a whole; seen, dialectically, as a projection through the template of metaphysical archetypes.

creation: origination; physical or psychological arrangement; mind creates with purpose, matter without; creation means active production but also passive result; a creation will have been informed by force of mind and/or matter.

D

dialectical stack: stack of opposites; columnar expression of polarity; there are two kinds of stack - primary or non-vectored and secondary, vectored; primary (essential) stacks set (*Sat*) Unity against ($\downarrow$ *tam*/ *raj* $\uparrow$) duality (for elaboration see especially Chapter 2 and *figs*. 1.4, 2.2, 24.1 and 24.2); secondary (existential) stacks represent the various kinds of polarity from which the changeful web of existence is composed; each pair of polar 'anchor-points' implies a scale or dynamic range that runs between 'paired opposition' or 'complementary covalency'; stacks do not necessarily list synonyms or make equations; ***their perusal is intended to promote connections because consideration of connections tends to help unify/ collate/ organise one's working comprehension of any matter in hand.***

DNA: a complex chemical; a large bio-molecule made of smaller units, nucleotides, strung together in a row; a polymer in the form of a double-stranded helix; a medium superbly suited to the storage and replication of 'the book of life'; 'paper and ink' on which the genetic code is inscribed; an organism's 'hard drive'; *DNA* is such an elegant, efficient and densely-packing form of information storage and manipulation that *DNA* computing by humans is now a fledgeling technology; in this fast developing field silicon-based technologies are replaced using *DNA* and other bio-molecular hardware.

E

electromagnetism: physics of the field that exerts an electromagnetic force on all charged particles and is in turn affected by such particles: light/ e-m radiation is an oscillatory disturbance (or wave) propagated through this field; light; light paradoxically involves a perfect, polar balance between contractive/ magnetic and radiant/ electric components.

elementary particles: science has discovered and, for the most part, experimentally verified, over fifty elementary particles; these are divided, in simple terms, into bosons (force carrying particles) and fermions (separate particles); bosons include photons (which mediate the electromagnetic force), gluons (which mediate the strong nuclear force), W and Z particles (which mediate the weak nuclear force), possibly gravitons (which mediate the gravitational force) and also possibly a Higgs boson (which may mediate a proposed mass-giving field); fermions include two main groups - six quarks and leptons (six electron/ neutrino types); derived from quarks are stronglyinteractive composites called hadrons; hadrons include baryons such as protons and neutrons and (perhaps a little confusingly) bosons such as short-lived mesons.

entropy:	a measure of the amount of energy unavailable for work or degree of configurative disorder in a physical system (see second law of thermodynamics); inertial aspect of an energetic, material or conscious gradient; diffusion or concentration gradient outward from source to sink; drop towards 'most probable' outcome i.e. inertial slack; a measure of disintegration or randomness; expression of the (*tam* ↓) downward cosmic fundamental; a major property of matter, closely coupled with materialisation; in a closed system, which the universe may or may not be, this tends the eventual loss of all available energy, maximum disorder and the exhaustion of so-called 'heat death'.
enzyme:	protein catalyst without whose type metabolism (and therefore biological life) could not happen.
epigeny:	genetic super-coding; contextual punctuation; chemical modification of *DNA*; also extra-nuclear factors that may cross- reference with genetic expression.
equilibrium:	three modes of equilibrium are (*sat*) balance of poise or pre-active potential; (*raj*) dynamic balance occurring in all regular cycles, wave-forms and cybernetic homeostasis that is basic to the stability of life-forms; and (*tam*) inertial equilibrium that results from diffusion of information or energy; it equates with exhausted inaction or 'flat', impotent rest; such post-active inertia represents the most probable distribution of energy/ matter with the least energy available for work viz. the most random arrangement permitted by the constraints of a system; expressed in psychological terms as ignorance, unconsciousness or sleep; see also equilibration, *karma* and *fig.* 3.3 'Pivoted Existence'.
Essence:	(*Sat*) Supreme or Infinite Being; Substance (perhaps Spinoza's Substance) 'prior to' or 'above' existence; Pure Consciousness/ Life; Peace that transcends all psychological and physical action; the root of an essentially undivided universe; Uncreated One within which and whence all differences have their being; Apex of Mount Universe; goal of saints/ 'philosopher kings'; the 'point' at which All-Is-One.
eukaryote	non-prokaryote; any organism except bacteria and blue-green algae.
evolution:	there are today *four* main usages of this word; each 'loading' derives from the original Latin, 'evolvere', meaning to unroll, disentangle or disclose; the *first two*, physical and biological, are conceived as natural/ mindless processes; the *second, mindful pair* is of psychological/ teleological import; specious ambiguity may conflate or switch between the fundamentally separate pairs of meaning. *Firstly,* in the scientific context of physics and chemistry, the word is used to describe change occurring to

physical systems; the laws of nature can't, it seems, evolve through time but stars, fires, rocks or gases can. *Secondly*, though also subject to the 'rules' of entropy, biological evolution is a theory of *random progression* from simple to complex form; it thereby implies increasing, codified complexity; while retaining the 'hard loading' of physical science it also, ambiguously, claims that codes, programs, mechanisms and coherent, purposive systems - normally the province of mental concept - self-organise by, essentially, chance; such confusion, the basis of naturalism, is compounded by failure to distinguish between, on the one hand, ubiquitously observed variation (called micro-evolution) and, on the other, Darwinian 'transformation' between different sets of body plan, physiological routines and associated types of organism - such 'black-box macro-evolution' as is never indisputably observed; to evoke a naturalistic ambience it is fashionable to use 'evolved' interchangeably with or to replace the words 'was created', 'was planned' or 'designed'; finally, it is noted that the coded, choreographed development of a zygote, packed with anticipatory information, through precise algorithms to adult form is the absolute antithesis of blind Darwinian evolution. *Thirdly*, man certainly evolves ideas; intellect can evolve 'purposive complexity'; we invent all kinds of codes, schemes and machines; we devise increasingly complex theories and technologies; and we evolve an understanding of natural principles; this, which all parties accept, is an informative, psychological sense of 'evolution'. The *fourth* sense of evolution, at least as near to the original Latin as the other three, is the spiritual usage; immaterial spiritual evolution, unacceptable to materialists and unknown to physical science, is at the very heart of holism; in this voluntary sense of evolution practitioners cast off material attachment, evolve and merge into the *Logos*; evolution (or, perhaps better, centripetal involution) of the soul is their great business; their aspiration is to unite with The Heart of Nature.

evolution pre-Darwinian: minority/ anti-mainstream pre-Socratic snippets and sense-based Epicureanism lionized by interpretations of post-18th century materialists; virtually undetectable eccentricity in Chinese, Indian and Islamic literature; natural selection treated by creationists al-Jahiz and Edward Blyth; Buffon, a non-evolutionist, addressed 'evolutionary problems'; Lamarck (evolution by inheritance of acquired characteristics); hints in poem by Erasmus Darwin.

evolution Darwinian: mechanism - natural selection; major tenets - common descent (inheritance), homology and 'tree of life'.

evolution: neo-Darwinian/ synthetic: as Darwinian, except synthetic theory adds random mutation as the mechanism for innovation; also adds a mathematical treatment of population genetics and various elements (e.g. geno-centric perspective) derived from molecular biology.

evolution: post-synthetic phase: natural selection and random mutation are acknowledged as mechanisms insufficient to source bio-information; post-Darwinian evolution invokes mechanisms from hypotheses such as *NGE* (natural genetic engineering) and 'evo-devo'; holistic possibilities also address the origin of complex, specified and functional bio-information.

existence: which 'stands out' from background 'nothingness'; the apparently divided universe; seemingly disparate, finite things; all motion/ change/ relativity; all psychological and physical events.

exon: specifies the amino acid sequence for a protein; m-*RNA* after protein editors have removed introns.

F

field: any extent wherein action either physical or metaphysical but of a certain kind occurs e.g. field of battle, influence of mind or magnetism; the scientific definition is limited to a collection of numbers varying from point to point - such as a scalar field of contours on a map - or numbers with direction - such as a vector field showing speeds and directions of wind.

first cause(s): first cause is first motion in a previously undisturbed, pre-conditional field; normally considered the first impetus to a chain of events; such 'horizontal causation' is complemented by the 'vertical causation' of Natural Dialectic's hierarchical view of cosmos as explained in Lecture 2.

First Cause Psychological is Archetype, Potential Informant or (see Chapter 5: Top Teleology) *Logos*; attributes of this Primary Source and Sustenance of Creation include omnipresence, omnipotence and omniscience.

first cause physical is called potential matter or archetypal memory; as the secondary source of creation it precedes physical phenomena; as such it is, transcending physical appearances, metaphysical; this 'physical nothingness' is therefore, paradoxically, the source of everything composing astronomical cosmos; it consists of their internal being as opposed to external manifestation or their essence as opposed to quantum or bulk state appearances; its void, with respect to the presence of finite phenomena, appears infinite; 'holographic' attributes of immanent archetype, the primary informant of our non-conscious, energetic universe,

137

include omnipresence and omnipotence; also check the Glossary for archetype and transcendent projection.

___

G

gamete: sex cell with half of full genetic complement i.e. a single set of chromosomes.

gene: generally means a basic unit of material inheritance; section of chromosome coding for a protein; digital file; a reading frame that includes exons and introns; the old one gene-one protein hypothesis is incorrect; in fact, by gene splicing, a particular piece of *DNA* may be used to create multiple proteins.

genome: total genetic information found in a cell: think of the genome as an instruction manual for the construction and physical operation of a given organism

genotype: the genetic constitution of an organism, often referring to a specific pair of alleles; the prior information, potential, plan or cause of an effect called phenotype.

gravity: in physics an attractive mass-to-mass force or warping of space-time; in Natural Dialectic the term is redefined more broadly - the agency of its (*tam* ↓) downward vector includes all psychological and physical factors of materialisation; such 'gravitational' factors and their properties are listed in the left-hand column of Secondary, Existential Dialectic; they include pain, pressure, confinement, strong nuclear force, mass, electromagnetic binding, inertia, entropy, 'standard' gravity and so on; gravity might be summarised as 'negative power' or 'the principle of death'.

___

H

holism: opposite of reductionism; the view that a whole is greater than the sum of its parts; the extra metaphysical (immaterial) ingredient is identified by Natural Dialectic as information; information implies the purposeful design, development and arrangement of contingent parts in a working system; cosmos may operate according to a Logical Norm.

hologram: A 3-d photograph made with the help of lasers. Unlike a normal photographic image each part of it contains the image held by the whole.

homeostasis: vibratory or periodic control of a system to obtain balance round a pre-set norm; the mechanism of its information loop involves sensor, processor and executor; the operative cycle works by negative feedback; psychological (nervous) and biological cybernetics; the informed basis of biological stability.

homeotic gene: gene involved in developmental sequence and pattern; high-level co-determinant of the formation of body parts.

138

I

illusion: is the cut between illusion and delusion an illusion? illusions, apparently outside the mind, appear real; a delusion, in it, we think real; illusion is a lesser truth; set against Absolute Truth (or Reality or Knowledge) it is a relative truth; hierarchical existence is composed of relative truths that range from slightly to completely false; only in Truth, only from the perspective of Knowledge do the illusions and delusions of existence wholly disappear (see Glossary truth; also *SAS* Chapter 4 and *PGND* Chapter 16: Truth, Appearance and Reality).

information: the immaterial, subjective element; information is action's precedent; informative potential is both a psychological and (by way of archetype) physical entrainer; information is the inhabitant of its own centre, mind, whose substrate is consciousness; active information knows, feels, purposes and codifies; it recognises meaning; on the other hand passive information reflects active; it is stored as subconscious memory; or is fixed in the expressions of non-conscious matter according, universally, to the archetypal behaviours of natural bodies or, locally, to particular constructions by life-forms.

informative entropy: loss of information due to degradation of its carrying medium; such a medium may be metaphysical (mind) or passive and physical (for example, computer files or genetic code); and its entropy may be metaphysical (loss of memory, focus or consciousness) or physical (for example, genetic mutation); the informative correlate of such degeneration is diminished organisational capacity, meaning or thrust of original purpose.

informative negentropy: gain of informative clarity; increasingly focused, purposive specificity; associated with knowledge, wisdom, grasp of principle and pristine construction.

intron: genetic control panel; n-p-c (non-protein-coding) segment(s) spliced from an m-*RNA* transcript prior to translation; introns include regulatory elements (to variably promote or inhibit gene expression) and addressing factors of the genetic operating system; gene-attached information lending specific flexibility to protein manufacture.

inversion: turning upside-down or inside-out; reversing an order, position or relationship; in a hierarchical sense inversion is allied with the reflective asymmetry of opposite poles; information outwardly expressed; pole-to-pole reversal integral to dialectical structure; various kinds of inversion (cosmic and micro-cosmic (or biological)) are discussed.

K

karma: the law of balance between action and reaction; equilibration such as underlies all mathematical equation; a deed with implications of the reactions or 'payback' it provokes; fruit or result of previous thoughts, words and deeds; applies as rigidly to metaphysical (psychological) as, in Newton's Third Law of Motion and mathematical *equations*, physical events.

L

levity: agency of the (*raj* ↑) upward vector; dialectical converse of gravity; psychological and physical 'levitatory' forces lift or stimulate; they are listed in the right-hand column of Secondary, Existential Dialectic and include light, heat, excitement, dematerialisation, release, negentropy, focus of interest, affection and so on; physically, levity includes anti-gravity or the intrinsic property of matter's absence, space; generally summarised as 'positive power' or 'a buoyant principle of liveliness'.

logic: analysis of a chain of reasoning; principles used in circuitry design and computer programming; 'normative reason' relates to the basic axiom(s) of a given standard e.g. *bottom-up* materialism or *top-down* holism; three main logical thrusts are: (1) inductive (premises/ observations supply evidence for a probable/ plausible conclusion) as in the case of experimental science working *bottom-up* from specific instances to general principle: (2) abductive (best inference concerning an historical event): and (3) deductive (conclusion in specific cases reached *top-down* from general principle): two pillars of logic are holism and materialism; holism employs mainly deductive/ abductive operations and a Logical Norm; materialism tends to inductive/ abductive operations whose axis is non-conscious force and chance.

Logos: First Cause; Prime Mover; Causal Motion that sustains creation's conscio-material gradient.

M

macrocosm: the physical universe of astronomy and cosmology; dialectically, the whole of existence (i.e. both universal mind and universal body) as opposed to individual, microcosmic objects and events - including the human body.

macro-evolution: large-scale, non-trivial evolution; process of common or phylogenetic descent alleged to occur between biological orders, classes, phyla and domains; includes the origin of body plans, coordinated systems, organs, tissues and cell types; unexplained by mutation, saltation, orthogenesis or any known biological mechanism; sometimes called 'general theory of evolution' (*GTE*); macro-evolution, an extrapolation from Darwinian micro-evolution vital to sustain the materialistic mind-set, is conjecture.

matter-in-practice: bulk, bonded matter including all molecular-based substances; gross matter; external appearance.

matter-in-principle: quantum phase; particles and forces; subtle matter; internal cosmic drivers.

meiosis: shuffling the information pack: variation-on-theme; mechanism for the production of haploid gametes; genetic postal system for sexual reproduction.

metabolism: body chemistry.

metaphysic: = non-physical/ immaterial/ psychological/ unnaturalistic (if physic is equated with natural); physically expressed as specific/ intended arrangement/ behaviour of materials; physical behaviour reflects metaphysical blueprint; involves element of information; also involves symbol/ code/ abstraction/ logic/ reason/ mathematics; also message/ meaning/ goal/ teleology; also consciousness/ mind/ life/ experience/ feeling; and also morality/ force psychological/ emotion; involves innovation/ creativity/ art/ invention/ aesthetics.

microcosm: an entity that reflects the universe by containing all its basic constituents. Used especially of the human state where it may refer to both mind and body or, in a purely physical context, body alone.

micro-evolution: misnomer; non-progressive, small-scale variation within a species or, more broadly, between strains, races, species and genera; *variation/ adaptation within type*; trivial Darwinian changes that may occur by natural selection/ ecological factors acting on genetic recombination or mutation; sometimes called 'special theory of evolution' (*STE*), micro-evolution/ variation is a fact.

mitochondrion: organelle in eukaryotic cells containing the apparatus for aerobic respiration.

mitosis: conservative copying and delivery of genomes in cell division; genetic reprinting; genetic postal system for asexual reproduction.

mnemone: a division of memory: the two divisions are personal mnemone (likened to a working cache or data store) and typical mnemone (likened to a *ROM* or an operating system); typical mnemone is a synonym for archetype; Natural Dialectic's definition (see especially Chapters 15 and 16) involves no 'cultural' connotation whatsoever and is thus wholly distinct from evolutionary psychology's use of the word; typical mnemone is a synonym for natural memory called, in any individual case, an archetype.

morphogene: part of typical mnemone or archetypal memory relating to physical construction; morphological attractor; the component of subconscious mind associated with electrochemical function and thereby body; morphogene is the dominant, perhaps exclusive, aspect of mind in unconscious organisms such as plants or fungi.

morphogenesis: the development of biological structure; more generally, the production of physical form.

mosaic: the presence of permutations of codified sub-routines or similarities of form and/or function scattered in organisms unrelated by lineage.

mutation: accidental change to genetic code.

mysticism: quite different from objective, it is the subjective science; not philosophy, religion or opinion but practice to achieve communion with natural, inner, immaterial truth; esoteric as opposed to exoteric, materialistic discipline; 'science of the soul'; as gyms and physical action are to athletes so meditative exercise and psychological stillness are to mystics; involves psychological techniques to achieve a clear, rational goal - purity of consciousness and thereby understanding of the fundamental nature of the informative principle, mind; since life is lived in mind a mystic seeks consummate knowledge of life's source and sanctum, that is, communion with its deathless heart; adepts were, are and will be 'Olympian' meditative concentrators.

N

nano-biology: biology of structures/ physiologies involving a few atoms or molecules; 'extremely small biology'.

nanotechnology: technology at atomic and sub-atomic level as is, basically, life's.

natural law: the automatic, reflex and mathematically describable behaviour of a physical entity; likewise the repetitive nature of its interactions with other entities.

naturalistic methodology: also known as 'methodological naturalism', this strategy is, strictly, not concerned with claims of what exists or might exist, simply with experimental methods of discovering physically measurable behaviours; thus only materialistic answers to any question (e.g. how biological forms arose or the nature of mind) are deemed 'scientific' or 'scientifically respectable'.

negentropy: opposite of entropy; lowering of entropy; expression of the (*raj*) upward-pointing cosmic fundamental closely coupled with stimulus, dissolution and dematerialisation; a measure of input, cooperation or synthesis; motive/ fluidising aspect of an energetic, material or conscious gradient; gain of energy, configurative order, information or consciousness in a system; when used in terms of information negentropy involves gain in order or understanding of principle from which different actualities derive; a measure of the amount of concentrated/ conceptual information, specific, intentional complexity or conscious arrangement in a system; a natural and essential property of mind.

nirvana: state of enlightenment; 'non-condition'; *nirvana* is devoid of existential motion; extinction of existence leaving Essence

Alone; pure soul; psychological super-state; Buddhists call such transcendence non-self or the Formless Self.

non-existence: where creation = formful existence, non-existence is formless; the polar opposite of physical space and time is Transcendent Potential; such pre- or super-existential formlessness is non-existent; Absolute Non-Existence is Essential; however relative non-existences of two kinds also occur; the first kind is metaphysical/ subjective and therefore psychological; it involves the absence of a specific psychological form or event; unconscious oblivion is one such non-existence; the second kind involves the local absence of a possible physical event (an object is a 'slow event'); impossibilities are non-existences but imaginations of non-existence (including symbolic abstractions, hypothetical entities, physical absences, absolute emptiness and the number zero) exist; furthermore, the nothingness of space and time, the zero-point of calculus and zero's empty set together constitute the basis of physical science and mathematics.

non-protein-coding *DNA*: occupies probably 95% of eukaryote and 80% of bacterial genomes; associated with the genetic operating system; may include some genuinely redundant misprints or duplications but now thought for the most part critical to the flexibility, efficiency and even possibility of gene expression; once thought of as useless, degraded information and ignorantly called 'junk *DNA*'.

non-protein-coding *RNA*: n-p-c *RNA* is also called nc-*RNA* (non-coding), nm-*RNA* (non-messenger) or f-*RNA* (functional); functional *RNA* molecule not translated into protein; many 1000's of different specimens include classes of t-*RNA* (transfer *RNA*), r-*RNA* (ribosomal *RNA*) and, commonly involved in the regulation of gene expression and other intra-cellular tasks, micro-*RNA*, double-stranded si-*RNA*, pi-*RNA* and so on; also, for inter-cellular communication, ex-*RNA*.

nucleic acid: *see DNA and RNA*

nucleotide: basic, triplex unit of nucleic acid polymer; monomer composed of phosphate and sugar (the 'paper' part) and base (the 'ink letter'); letters' of the genetic alphabet are (G) guanine, (C) cytosine, (A) adenine and (T) thymine. In *RNA* thymine is replaced by (U) uracil.

O

object: a slow, although energetic, event; apparent fixation.

Om: universal sound, fundamental reverberation, basic creative agent; sometimes spelt *Aum*, a Sanskrit word whose Semitic transliterations are Am'n, Amin and Amen; see also First Cause, *Logos*, *Kalam*, *Shabda* etc.

order: regular, regulated or systematic arrangement; organisation according to the direction of physical law; passive information by which things are arranged naturally (with predictable but non-purposive complexity) or purposely (with innovative or specified complexity); mind, generating specified complexity in the order of its technologies and codes, actively informs; the orders of mind are meaningful, the orders of matter lack intent.

organelle: cellular sub-station; discrete part of a cell; sub-cellular compartment having specific role such as informative (nucleus), energetic (mitochondrion, chloroplast), constructional (ribosome, Golgi body) or other.

P

PAM, *PAND*, *PCM* and *PCND*: philosophical gambits; see Primary Axioms and Corollaries.

phenotype: the effect of causal potential; result of the development of prior, informative 'egg'; outward expression of inner plan; sensible appearance of an organism as opposed to its genotypic scheme: the whole set of outward appearances of a cell, tissue, organ and organism are sometimes called a phenome (*cf.* genotype/ genome).

photosynthesis: process by which inorganic carbon is introduced to the biological zone and energetic sunlight fixed as a crystalline molecule of storage, a sugar called glucose.

phylogeny: evolutionary history; relationships based on common or evolutionary descent.

potential: poise; latent possibility; potent non-action that precedes any particular action or creation; in science potential energy is defined as the energy particles in a system (or field) possess by virtue of position/ arrangement; gravitational, electrical, electro-chemical, thermo-dynamical and other kinds of potential are recognised; in dialectical terms mind precedes matter, information precedes the pattern of material behaviour; information is energy's pre-requisite potential; in this case *informative potential* involves two conditions; firstly, a pre-existential/ essential state of pure potential; secondly, a pre-material, metaphysical fact of potential matter, archetype or laws of nature; if potential's pre-active equilibrium is related to the voltage of a full battery then aspects of psychological 'voltage', whose currents drive intentional behaviour, are purpose, will and plan.

potential matter: see archetype.

Primary Axiom of Materialism (*PAM*): all objects and events, including an origin of the universe and the nature of mind, are material alone; cosmos issued out of nothing; life's an inconsequent coincidence, a fluky flicker in a lifeless, dark eternity.

Primary Axiom of Natural Dialectic (*PAND*): material derives from immaterial; a conscio-material dipole that issues from First Cause informs and substantiates both mental

144

(metaphysical) and physical creations; there is eternal brilliance whose shadow-show's creation.

Primary Corollary of Materialism (*PCM*): the neo-Darwinian theory of evolution, that is, life forms are the product, by common descent, of a random generator (mutation) acted on by a filter called natural selection; such evolution is an absolutely mindless, purposeless process; the *PCM* is a fundamental *mantra* of materialism.

Primary Corollary of Natural Dialectic (*PCND*): the origin of irreducible, biological complexity is not an accumulation of 'lucky' accidents constrained by natural law and death; forms of life are conceptual; they are, like any creation of mind, the product of purpose.

prokaryote: non-eukaryote; bacterial type with little or no compartmentalisation of cell functionaries.

promissory materialism: belief sustained by faith that scientific discoveries will in the future justify/ vindicate exclusive materialism and, as a consequence, atheism; may involve a call to progress towards the technological provision of its 'promised land'.

protein: factor made from a specific sequence of amino acids to perform a specific task; 'informative' protein includes some hormones; skin, hair, bone, muscle and other tissues are made of 'structural' protein; 'functional protein' called enzymes mediates all stages in cell metabolism, that is, it catalyses all biochemistry.

***PSI* (psychosomatic interface)**: psychosomatic border; the level of mind-matter interaction; bridge between metaphysical and physical dimensions; potential matter; 'gap of Leibniz'; 'fit' of mind to matter; point of linkage between subconscious mind and non-conscious matter; gearing between instinct/ archetype and the behaviour of material objects and energies; as in the case of physical law, psychosomatic influence is both general in potential and local/ specific in engagement.

psychological entropy: a measure of loss of concentration, focus of attention or consciousness; loss of 'mental energy' or aptitude; the drop from waking to sleep; loss of knowledge, information or sensitivity; the gradient from intelligence through stupidity to oblivion; an expression of the (*tam*) downward cosmic fundamental in mind; a tendency predominant in lower, egotistical or selfish mind; increasing level of ignorance, anguish or immorality; loss of integrity, psychological disharmony or disintegration; see also *information entropy*.

psychological negentropy: a measure of gain in order; an increase in concentration, focus of attention or consciousness; gain in sense of purpose, 'mental energy' or aptitude; the rise from sleep to waking, 'dark to light' or unhappiness to happiness; gain in knowledge, information or sensitivity; the gradient of learning and spiritual evolution; an expression of the (*raj*) upward cosmic fundamental in mind; a tendency predominant in higher mind; increasing level of contentment, understanding and the natural morality of happiness; the ascent towards psychological radiance, harmony and integration. The converse of psychological negentropy involves *entropy of information.*

psychosomasis: operation across the psychosomatic border; mind/ body interaction; the one-way, morphogenic imposition of archetypal pattern on *physicalia*; the two-way exchange of information in sentient organisms through the agency/ medium of subconscious patterns.

Q

quantum: minimum discrete amount of some physical property such as energy, space or time that a system can possess; quantum theory states that energy exists in tiny, discontinuous packets each of which is called a quantum; an elementary discontinuity; an elementary particle e.g. photon or electron.

quantum level: matter-in-principle; 'internal', 'causal' or 'subtle' matter; the vibrant or energetic phase of physical organisation; zone of sub-atomic particles and forces; step (on cosmic ziggurat) between potential and bulk matter whose aspect is sometimes extended to include atomic and molecular interactions; small-scale substance underlying large-scale, sensible appearances.

R

raj: (↑) upward, levitatory or stimulatory cosmic vector.

reductionism: opposite of holism; the materialistic view that an article can always be analysed, split up or 'reduced' to more fundamental parts; these parts can then be added back to reconstruct the whole; a whole is no more than the sum of its parts.

religion: etymology debated between Latin *relegere* (review) and *religare* (bind); this latter is the sense of *yoga* which also means joining, yoking or connection; *religio* means dutiful and meticulous observance; currently religion means world-view, mind-set or basic faith; whether of materialistic or holistic belief, it involves the non-negotiable substance of an individual or community's truth - notably as regards origins; antagonism between holistic practice and the naturalistic methodology of science is, because the couple deal with separate but complementary physical and metaphysical dimensions, flawed; a materialist/ atheist 'binds meticulously' to an evolutionary mind-set, a holist to either pantheism or a Primary Living Creator; in the case that self-deception is crucial to successfully deceiving others which, holism or materialism, is the religion that is ultimately true?

resonance: the tendency of a body or system to oscillate with a larger amplitude when subjected to disturbance by the same frequencies as its own natural ones; thus a resonator is a device that naturally oscillates at such (resonant) frequencies with greater amplitude than at others; resonance phenomena occur with all kinds of vibration, oscillation or wave; their sorts include mechanical, harmonic (acoustic), electrical (as with antennae), atomic and molecular.

respiration: the controlled release of energy from food.

ribosome: site of polypeptide (protein) synthesis.

RNA: a single-stranded nucleic acid polymer employed in three different forms during the process of protein synthesis; in computer terms might be likened to a portable memory stick as opposed to *DNA*'s hard drive.

S

sanskara: character trait; groove, habit, obsession or repetitious mode of thought proportional in depth to the intensity of desire, force of impact or impression that created or sustains it.

sat: 'top' or essential cosmic fundamental; 'vector' of balance, neutrality.

science: Latin *scire* (know); knowledge; commonly understood as the practical and mathematical study of material phenomena whose purpose is to produce useful models of the physical world's reality.

scientism: a philosophical face of official, *de facto* commitment to materialism; today's majority consensus of what the creed of science is; an -ism born of *PAM*; a faith that all processes must be ultimately explicable in terms of physical processes alone; like communism, a one-party state of mind; a doctrine that physical science with its scientific method is ultimately the sole authority and arbiter of truth; a set of concepts designed to produce exclusively material explanations for every aspect of existence, that is, to colonise each academic discipline and build its intellectual empire everywhere; 'scientific fundamentalism' closely allied, when expressed in social and political terms, with 'secular fundamentalism', sociological interpretation of behaviour and the fostering of a humanistic curriculum.

secular fundamentalism: *PAM* as applied to the worlds of nature and of human society.

secularism: concern with worldly business; lack of involvement in religion or faith; secularism is generally identified, as defined by the dictionary, with materialism; for a secularist the ultimate arbiter of truth is human reason - ideas are open to negotiation so that even morality is relative;

however many liberal agnostics, atheists and humanists argue that their metaphysical, philosophical system also embraces so-called 'universal' moral values and, as opposed to zealotry or the logic of evolutionary faith, a liberal politic of 'philosophical live-and-let-live'.

sub-state: *opp.* super-state; impotence, discharge, exhaustion, final stage in the expression of potential; fixity; non-conscious base-state; state 'below/ subtendence; extreme negativity/ (*tam*) condition.

super-state: potential; source of possibility; causal metaphysic/ archetype; state 'before' or 'above' subsequent expression; immanence; transcendence; precondition; (*sat*) priority.

symmetry: an aspect of the (*sat*) characteristic of balance; aesthetically pleasing balance and proportion; geometrical mirror image or interactive process such that some feature of an action remains invariant, that is, conserved; the symmetry of an entity (such as a sphere, empty space or natural law) or operator (such as energy) that remains the same at all times everywhere from any local point of observation or through every transformation is called 'higher' or 'continuous'; if a feature is conserved only when an object or process is moved, turned or viewed at certain angles or under specific conditions its symmetry is called 'lower' or 'discrete'; the symmetrical properties of a system may be precisely related to corresponding conservation laws and *vice versa*; internal symmetries found in quantum physics (such as gauge transformations) are independent of space-time coordinates; scale symmetry occurs when a reduced or expanded object keeps its shape but not its size (as with Mandelbrot fractals); dialectical symmetry involves the balancing of complementary opposites (as exemplified in the yin-yang symbol); it also involves *informative potential*; its metaphysical archetypes inform principles, laws or determinant fields that exist prior to action and, from their possibilities, govern actual outcome; such 'configuration of the world' is absolute and, beyond entropy, stable; it is negentropically immune from decay; by contrast, the 'free' symmetry of *potential energy* is inherently unstable and (like a pencil balanced on its tip) liable to spontaneously 'topple' or 'break' into the least energetic of a range of circumstantial possibilities; such spontaneous symmetry-breaking, the basis of diversity, represents an expression of 'deep symmetry' or archetype under local conditions and is therefore called by physicists 'contingent'.

T

tam: (↓) downward, gravitational or inertialising cosmic vector.

teleology: the doctrine that there is evidence of purpose in nature; doctrine of non-randomness in natural architecture; doctrine of reason ('for the sake of', 'in order to', 'so that' etc.) and intent behind biological and universal design.

third eye: place where you think; point of metaphysical focus between and behind the eyebrows, that is, just above the physical eyes; HQ/ seat of mind beyond the sensory world; cosmological eye-centre; gate through which meditative concentration can pass; single way that leads within.

transcendent projection: creative projection from intrinsic Source; see Primary (Metaphysical) and Secondary (Physical) First Causes from Lectures 3 and 4; voluntary issue from potential mind or involuntary from potential matter; in other words, transcendent metaphysical projection is through Causal Archetype and transcendent physical projection through mnemone (see Index for references).

psychological: projection is through Causal Archetype called *Logos*, Holy Name, *Sat Nam*, *Om* and many other names. This projection is from Alpha Point, the first and highest cause of creation; see also *SAS* Chapter 5: Top Teleology and *figs. PGND* Chapter 13: First State of Super-consciousness.

physical: as Lecture 4 explains, physical first cause is projection through potential matter or base archetype in universal mind; such memory is the source of cosmo-logical language; it involves the orderly expression of energetic forces and particles; linkage, as Lecture 3 suggests, by psychosomatic resonance; emergence of physical phenomena from archetypal noumenon, that is, from unseen potential; an instantaneous 'miracle' that issues from 'within' non-conscious physicality; transcendently emergent, finely tuned expansion from 'inner' metaphysic into 'outer' material/ natural law; physical nothingness from whose from whose prior pointlessness (or 0-dimensional singularity) all points began; cosmic seed whence, *ex nihilo*, the world developed; projection whose appearance, once physical, is visible and perhaps described but certainly not explained by big bang theory; transcendent projection of archetype is possibly, to the constrained sensory and intellectual states of human mind, ultimately incomprehensible; its invisible dynamic, the practice of materialisation, may remain a fact beyond material understanding. for references involving more detail about psychological, physical and biological projections see *SAS* or *PGND* Glossaries.

biological: if matter is developed memory (*PGND* Chapter 14: Space) then Lectures 3 and 5, *SAS* Chapters 16 and 19 and *PGND* Chapters 13: Typical Mnemone and 15: Conceptual Biology expand this theme; biology is based on information in the form of code and codification, demanding forethought, is a product of mind; intrinsic archetypal program is reflected by extrinisic chemistry called *DNA*; in terms of Natural Dialectic *DNA* is passive information.

truth: what's correct or accurate; a universal truth substantiates, as source and sustenance, all things; man's holy grail is truth; *top-down*, within a hierarchically devolved construction (such as a machine, cosmos or other working system) truth is vested both in its source and physical appearance; in this case, viewed at their own level, psychological (subjective) or physical (objective) facts appear self-evident, obvious realities; however, they are relative or lesser truths when set against Truth Absolute; from different heights on a mountain slope the climber's perspective changes; so with Mount Universe, whose Peak represents Whole Truth; in the light of Peak Reality all existence seems real but is composed of relative illusions; it is a shadow play dependent on The Independent Source; man is born to seek and find this Essential Source, this Alpha Point, his Origin; see also Glossary: illusion.

U

unification: details are unified by their working principles, themes or programs; better to perceive intrinsic principle is to simplify or unify an understanding; progressive unification of forces is the grail of physics: Clerk Maxwell unified electricity and magnetism; electroweak or *GSW* theory brought in the weak nuclear force; now the goal is to include the strong nuclear force (*GUT*), gravity in a super-force and show that, in essence, particles and forces are interchangeable (super-symmetry and *TOE*); Natural Dialectic, also working with the maxim 'All is One', includes what sums to a hierarchical *TOP* or Theory of Potential; potential (see *SAS*: Glossary and Index for archetype and potential) is the absolute from which variant orders of relativity derive; the subjective potential for mind is consciousness and the objective potential for matter is archetypal memory; such archetypal element unites psychology with the physics of natural science; it is the informative precondition of physical and biological form.

V

virtuality: exotic component of quantum physics; para-physical feature of the quantum vacuum; immaterial substrate of material phenomena; practical metaphysic; inner (where solidity's the outer) edge of physical reality; ephemeral 'virtual particles' rise and sink back into a 'void' thought to teem with their 'fluctuations'; virtuality is identified as the agent of such important actualities as the strong nuclear force (resulting from interaction between virtual mesons and gluons), vacuum polarisation, the Coulomb force (between electric charges and mediated by the exchange flight of virtual photons) and so on; not used here in the computer sense of a 'virtual' continuum between real and imaginary circumstance; see also *ZPE*.

Z

zero: zero (the number) is a metaphysical entity, one critical to mathematics; zero (the fact) means, for Natural Dialectic, nothing in two senses; in the *negative sense* it means an absence of perception (psychological oblivion) or absolutely nothing physical (as naturalistically prescribed to precede, say, a big bang or as the nature of a theoretically perfect vacuum); negative sense may also be construed as (*tam*) an extreme sub-state, sink or emptiness; for materialism 'absolute nothingness' may involve natural law and its mathematical description; what, one may enquire, is the source of such 'eternal metaphysic', what is the nature zero-physical?: on the other hand, in a *positive sense* zero refers to source, pre-existent potential or (*sat*) higher cause-in-principle; for example, information (which is zero-physical) transcends/ precedes a course of action; information that passively governs the operation of cosmos derives from immaterial archetype.

ZPE: zero-point energy; quantum vacuum; vacuum energy of all fields in space; residual energy of all oscillators at $0°K$; concept first developed by Albert Einstein and Otto Stern; intrinsic energy of vacuum; the ground-state minimum that any quantum mechanical system, in particular the vacuum, can have; remainder, according to the uncertainty principle, when all particles and thermal radiation have been extracted from a volume of space; residual non-thermal radiation; irreducible 'background noise'; 'quantum foam'; the potent, microscopic side of quantum vacuum (as opposed to impotent, macroscopic vacuum left by the apparent lack of anything); subliminal 'rumblings' of immaterial weak, strong and electromagnetic fields (called *ZPF*s); seething, jostling ferment of subliminal waves and particles in emptiness; a flux of unobservable 'virtual' matter and anti-matter that may or may not appear as the basis of observable forces such as electromagnetism, charge and perhaps inertial mass and gravity; a subtle facet of levity; the anti-gravity of dark energy (or the cosmological constant) has been postulated as a component of *ZPE*; suggested 'mother-field' support for electron orbits, atomic structure and thus the phenomenal universe.

zygote: fertilised egg.

Connections

SAS   Science and the Soul
A&E   Adam and Evolution
AMA   A Mutant Ape? The Origin of Man's Descent
PGND A Potted Grammar of Natural Dialectic

---

1   *SAS*: Chapter 0: Anti-parallel Perspectives; also *PGND* Chapter 1: Primary Assumptions; *A&E*: Chapter 2.
2   *PGND* Chapter 1: Primary Assumptions.
3   *SAS* Preface; *PGND*: Chapter 2.
4   *SAS*: Chapter 1 Natural Dialectic's ABC; *PGND*: Chapter 3 Models.
5   *SAS*: Chapter 1 Natural Dialectic's ABC/ Glossary/ Index; *PGND* Chapter 4.
6   *SAS*: Chapter 1 Natural Dialectic's ABC/ Index: stack and dialectical operator; *PGND* Chapter 5/ Index: stack and dialectical factor.
7   *SAS* fig. 2.4.
8   *PGND* Chapters 13 to 15.
9   *SAS*: Chapter 0; *PGND* Chapter 1: Two Pillars, A Dialogue of Faith.
10 *SAS* Chapter 2: First Principles; *PGND* Chapter 10: Existence.
11 *SAS* Chapter 3 and *PGND* Chapter 7: A Hierarchical Perspective; *A&E*: Chapter 3.
12 *SAS* Chapter 5 and *PGND* Chapter 11.
13 *SAS* Chapter 5: (*Sat*) Potential Information and Chapter 6; *PGND* Chapter 12.
14 *SAS* Chapter 6: Information's Infrastructure - Code; also Glossary and Index; *PGND* Chapter 12.
15 *SAS* Chapter 6: Music.
16 *SAS* Chapter 6: Machines, also Mind Machines (Computers).
17 *SAS* Chapter 13: Psyche and Psychology.
18 *SAS* Chapter 13: The Neurological Delusion.
19 *SAS* Chapter 13: Consciousness; *PGND* Chapter 13.
20 *SAS* Chapters 13, 14 and 15; *PGND* Chapter 13: Triplex Psychology.
21 *SAS* Chapter 14: Second and Third States of Consciousness.
22 *SAS figs*. 13.2 and 13.3.
23 *SAS* Chapter 13: Build Yourself a Brain.
24 *SAS fig*. 13.4.
25 *SAS* Chapter 24: Twists in the Bio-logical Tale.
26 *SAS* Chapters 15 and 16; *PGND* Chapter 13: Third Sate of (Sub-) Consciousness.
27 *SAS* Chapters 15 and 16: Personal and Typical Mnemones; *PGND* Chapter 13: The Typical Mnemone and *H. archetypalis*; Glossary and Index of both books.
28 *SAS* Chapters 16, 17 and 19; *PGND* Chapters 13 and 15; Glossary and Index in these and also in *A&E* and *AMA*.
29 *SAS* Chapters 15 and 16: Psychosomatic Linkage and Psychosomasis; *PGND* Chapter 13.
30 *SAS fig*. 15.4.
31 *SAS fig*. 15.5.
32 *SAS* Chapter 16; *PGND* Chapter 13: *H. electromagneticus*.
33 *SAS* Chapter 16: Synchromesh 2 - Psychosomasis; *PGND* Chapter 13 last

three sections; see also Glossary: resonance and indices: resonance, harmonic oscillation and vibration.

[34] *SAS* Chapter 7: Lady Luck and Lord Deliberate.

[35] *SAS* Chapter 9.

[36] *SAS* Chapter 10.

[37] *SAS* Chapter 11: Alpha Points.

[38] *SAS* Chapter 12: Magnificent Mythology.

[39] See Lecture 2: (*Sat*) Potential Information; also this chapter First Cause Psychological; *SAS* Chapter 5: Top Teleology.

[40] *SAS* Chapter 20; *SAS* Glossay and Index: code; *A&E* Chapter 4: Genes and Genesis.

[41] *SAS* Chapter 23: Super-codes and Adaptive Potential

[42] *SAS* Chapters 5, 6 and 19; *PGND* Chapter 15.

[43] *SAS* Chapter 19: The Central Executive is Homeostasis; Index: balance and equilibrium; see also *PGND* Chapter 15.

[44] *SAS* Chapter 21: Energy Metabolism Perchance?; *A&E* Chapter 10: Life's Engine.

[45] *SAS* Chapter 19: Nuclear Super-computing and Conceptual Biology.

[46] *SAS* Chapter 22: *A&E* Chapter 3: Hierarchy.

[47] *SAS* Chapters 20 and 21; *PGND* Chapter 15: Chemical Evolution?; *A&E* Chapters 4, 8, 9 and 10.

[48] *SAS* Chapter 22: The Editor; A&E Chapter 5: Sports, Survival and the Hone.

[49] *SAS* Chapter 23: The Creator; *A&E* Chapter 5.

[50] *SAS* Chapters 21 and 23; *A&E* Chapters 2 and 9.

[51] also *SAS:* Chapters 7 and 20: Numbers Games; and *PGND* Chapter 15 footnote 150 concerning Wistar.

[52] In the 'Origin of Bio-information and the higher taxonomic categories' (Proceedings of the Biological Society of Washington 4-8-2004) S. Meyer argues that no mechanistic theory can account for the amount of information needed to build novel forms, systems or organs of life.

[53] *SAS* Chapters 24 and 25; *A&E* Chapters 7 and 8.

[54] *SAS* Chapter 24; *A&E*: Chapters 7 and 12.

[55] *SAS* Chapter 24; *A&E* Chapter 7.

[56] *SAS* Chapter 25: The Origin of Growth and Development.

[57] *SAS* Chapter 25: A Clap of Fragile Wings.

[58] *SAS* Chapter 26.

[59] Absolute Community of Essence and Existence: see also *SAS* Chapter 4 and *PGND* Chapter 16: Truth, Appearance and Reality.

[60] Relative Community of Existence; as with previous end-note.

[61] Ecology: *SAS* Chapter 26.

[62] Negativity involuntary (natural) and voluntary.

[63] Mein Kampf: Chapter 4.

[64] Unified Theory of Community: Social and Individual Parts: SAS Chapter 26.

[65] Individual Association: *SAS* 4, 26 and 27; *PGND* 16: Science to Conscience and Is There an Absolute Morality?

[66] Religion is a word derived from the Latin 'to bind back'; 'yoga', equally, means binding to (a discipline).

Index

power/ harmonic oscillation

Love

Brilliance/ *GUE*/ Unity/ Communion123

LUCA (Last Universal Common
Ancestor) 107

M

163

T

U

V

The author has recently written a few more books (available from Amazon, Foyles, Waterstones, Barnes & Noble etc. and see website addresses on p.2):

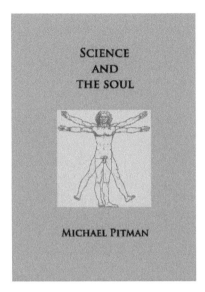

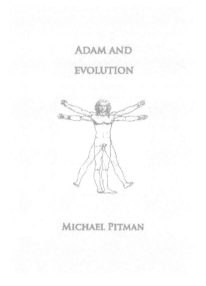

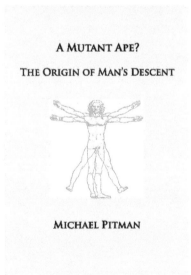

Illustration credits: Wikipedia or Wikimedia Commons; GNU Public Licence 3: Kalachakra Mandala: Wikimedia Commons; author Kosi Gramatikoff.

Lightning Source UK Ltd.
Milton Keynes UK
UKHW02f0220260118
316877UK00007B/43/P